COVID-19

T0203717

This book is devoted to the efforts of Environmental Health Practitioners (EHPs), their employers and supportive professional bodies world-wide in responding to the COVID-19 pandemic.

Drawing upon the first-hand experiences and reflections of EHPs working across the professional discipline in countries around the world, the book highlights how they responded to the initial wave of SARS-CoV-2 infection as it spread globally. It explores how this impacted on their environmental health work as their wider public health skills and expertise were increasingly called upon. The book recognises the significant contributions that EHPs have made to protect lives and livelihoods since the seriousness of COVID-19 became apparent. It also identifies shortcomings in the response and deployment of personnel and makes a series of recommendations to inform future practice.

This book:

- Captures a moment in history through the experiences of EHPs in meeting the complex challenges presented by the COVID-19 pandemic.
- Features the observations of front line practitioners on the practical challenges and opportunities encountered globally, suggesting the lessons learnt for current practice in infectious disease prevention and control.
- Expands upon the reflections of some of the professional bodies around the world as to how the response of EHPs to the COVID-19 pandemic should result in a renewed commitment to public health through Environmental Health.

EHPs in current practice and in training, other public health professionals and those looking to build better health protection services, now, and in the future, will find this book a valuable resource to inform the case for the key role of Environmental Health in the current pandemic, in response to future challenges and crises, and in managing risks to health encountered in more usual times.

Dr **Chris Day** is the editor and lead author, who began his career in 1975. Since then, he has worked in local government, provided consultancy services and training in industry and lectured for 17 years on the BSc Environmental Health course at King's College London, where he latterly led the MSc programme. His Doctorate in Public Health was obtained from the London School of Hygiene & Tropical Medicine, where his thesis focused on the surveillance of food-borne illness. In 2010 Chris joined the Chartered Institute of Environmental Health, where he was part of the team responsible for developing the 2011 Curriculum and professional training portfolio and editor of the *Journal of Environmental Health Research*. He has been involved in several funded research projects over the years and contributed chapters to recent editions of *Clay's Handbook of Environmental Health*. Although now semi-retired, Chris is still undertaking consultancy work.

Routledge Focus on Environmental Health

Series Editor: Stephen Battersby, MBE PhD, FCIEH, FRSPH

For more information about this series, please visit: https://www.routledge.com

COVID-19
The Global Environmental Health Experience

Chris Day

**With contributions from
Rob Couch and Surindar Dhesi**

Routledge
Taylor & Francis Group

LONDON AND NEW YORK

First published 2021
by Routledge
2 Park Square, Milton Park, Abingdon, Oxon OX14 4RN

and by Routledge
605 Third Avenue, New York, NY 10158

Routledge is an imprint of the Taylor & Francis Group, an informa business

British Library Cataloguing-in-Publication Data
A catalogue record for this book is available from the British Library

Library of Congress Cataloging-in-Publication Data
Names: Day, Chris, author.
Title: COVID-19 : the global environmental health
experience / Chris Day ; with contributions from Rob Couch
and Surindar Dhesi.
Description: Milton Park, Abingdon, Oxon ;
New York, NY : Routledge, 2021. |
Series: Routledge focus on environmental health |
Includes bibliographical references and index.
Identifiers: LCCN 2021011201 (print) |
LCCN 2021011202 (ebook) |
Subjects: LCSH: COVID-19 (Disease)—Environmental
aspects. | Environmental health. Classification:
LCC RA644.C67 D39 2021 (print) |
LCC RA644.C67 (ebook) | DDC 614.5/92414—dc23
LC record available at https://lccn.loc.gov/2021011201
LC ebook record available at https://lccn.loc.gov/2021011202

ISBN: 978-0-367-74316-1 (hbk)
ISBN: 978-0-367-74331-4 (pbk)
ISBN: 978-1-003-15722-9 (ebk)

DOI: 10.1201/9781003157229

Typeset in Goudy
by codeMantra

Contents

Foreword

Our global Environmental Health workforce serves to guard the environments in which people live, work, play, study and carry out the multitude of other activities that make for a happy and healthy life. This special group of health professionals know about the hazards that put lives at risk, and through working in the community, understand how to get their different 'messages' across.

With the COVID-19 pandemic, we are surely all now living through the most awful of times. While the virus itself is indifferent as to who it infects – and the level of infection worldwide is a terrible reminder of how vulnerable we all are – it manages to affect the poorest and weakest the worst. Of course, now, we are beginning to see 'the light at the end of the tunnel' – vaccination!

Universal vaccination has the ability to reduce suffering, save lives and return us to something like pre-pandemic normality. However, it can only do this if those manufacturing vaccines, and, more importantly, those controlling their availability and distribution, can be encouraged to work with emerging countries, so providing them with supplies, at scale, in time and at affordable cost. There must surely be something that Environmental Health and its practitioners could do to assist, whether it be developing policy, managing human resources, or seeing to it that materials are conveyed to the point of vaccination, safely, securely and hygienically.

At the same time, we need to give serious thought to how we got here, since in well-preserved habitats, with a wide diversity of species living in balance, the viruses that threaten our existence now would be distributed more widely among different species, and in that sense would be 'remote' from humans. This pandemic is not part of a unique event, but of a pattern that reflects the damage we are doing to the habitats where we all – plants, animals and humans – have a right to co-exist. However, as a consumerist society that needs more than that which nature can provide, we have exhausted its resources, and its patience. We must now seek to rediscover the objectives the world had set itself when sustainable development gave mankind the will to think globally and act locally.

It is now more important and urgent than ever that the world quickly understands the inter-connectedness of the environment, health and economy, or as Dr Tedros Adhanom Ghebreyesus, director general of the World Health Organization, put it recently, "Attempting to save money by neglecting environmental protection, emergency preparedness, health systems, and social safety nets, has proven to be a false economy – and the bill is now being paid many times over." It is time, then, to prepare and invest in a healthy and green recovery, close to all communities, with Environmental Health leading the collaboration through the International Federation of Environmental Health.

In the meantime, as SARS-CoV-2 threatens to do further harm, Environmental Health practitioners around the world are at the forefront of this struggle to contain the infection as the last line of defence. They have shown, in the year since the pandemic was declared, the importance of Environmental Health, and, through their efforts, the vital and courageous role they have played in seeing the various preventive measures to combat COVID-19 implemented.

Whether it be tracking contacts, enforcing government guidelines or sharing wisdom with their communities and colleagues from other disciplines, as this book demonstrates, they have been out there meeting the challenge. However, in looking further to discover the challenges encountered by EHPs in these early months of the pandemic, and, more importantly, the solutions found, and opportunities discovered, the book may serve to inform fellow practitioners and those in a position to decide on enhanced investment in Environmental Health.

Susana Paixão, BSc(Hons), MSc, PhD
*President, International Federation of Environmental Health, also
President of the Portuguese Society of Environmental Health and
Professor at Polytechnic Institute of Coimbra, Portugal, Coimbra Health
School, Environmental Health Department*

Preface

The COVID-19 pandemic demanded swift and decisive action on the part of a world that should have learnt its lessons from similar challenges over the past 20 years but didn't. That some countries did so only served to draw attention to the efficacy of public health measures when applied with gusto. For the rest, it would be – sadly still is – a horrible and seemingly remorseless game of 'catch-up' as they seek to protect their people against SARS-CoV-2 until biotechnology can weave its magic.

Aside from the selfless healthcare personnel who, knowing the risks, put themselves in harm's way, saved the lives of others and lost theirs, were a cadre of Public Health professionals called Environmental Health Practitioners. Working in all tiers of government, academe, the military and private sector, EHPs migrated from their 'routine' work of protecting lives at risk from unsafe food, hazardous workplaces and unsatisfactory housing to working increasingly on public health measures focussed on COVID-19.

Considering the first nine months of the pandemic, from the point when the WHO declared SARS-CoV-2 a 'Public Health Emergency of International Concern' on 30 January 2020, we sought to discover, among other things, what EHPs around the world had been doing, what special challenges they had encountered, and how they thought the discipline of Environmental Health might have been enhanced by their endeavours.

The outcome of this field enquiry was remarkable in that an otherwise retiring group of people, normally reticent when speaking of their skills and achievements, talked of rediscovering their professional roots and sensing they had saved lives. Their new-found roles sometimes owed more to the nature of the public health service before COVID-19, and the enlightenment (or otherwise) of their employers, but these ranged from leading contact tracing teams to overseeing the burial of the dead, and everything conceivable in-between.

The difficulties encountered along the way have clearly been many and various, and have, in some cases, been daunting and remain unresolved. However, the EHPs who shared their 'stories' with us seem to have tackled these obstacles head-on, or taken a pragmatic route around them. They often did this by joining up with other health professionals to develop strategies and initiatives to deal with circumstances playing out in real time. However, when situations demanded it, and they regularly did, they shunned the telephone and 'Zoom' and chose to go out 'on the district', for as one EHP put it, '... that is where the work really is'.

Above all, they brought to public health their knowledge, experience and deep understanding of how people live, work and otherwise 'tick'. I will close this preface with a comment received from an EHP in the United States:

> This profession is not glamorous; you are only quietly rewarded for your efforts. No one knows the work you do, which is why a lot of the people in this profession do it. We do not want to be the face of movement, just the gears and the minds to be out there protecting the public...
>
> Chris Day, MA, DrPH, MCIEH, CEnvH

Series preface

There are now seven publications in this relatively new series, and more are in the pipeline. The aim remains as ever the same – to explore Environmental Health topics, traditional or new, and raise sometimes contentious issues in more detail than what might be found in the usual Environmental Health texts. That is exemplified by this latest edition that looks at the global Environmental Health experience during the SARS-Cov-2 pandemic, and the varied work and activities of EHPs.

This series is an important part of the professional landscape, as is apparent from the titles published so far and those in the pipeline. Environmental Health practitioners bring their expertise to a range of situations and are deployed differently but not always to good effect so far as public health is concerned. It may be because in some countries, governments are unaware of what is Environmental Health, or practitioners have a 'low profile' and are taken for granted. This particular monograph rather highlights this. In the same way some states have been more effective than others in dealing with the pandemic and minimising infection rates and mortality (and have also suffered less economic damage). It is hoped that this series will be used as a means of highlighting the work of Environmental Health practitioners.

At the same time, we want to encourage readers and practitioners, particularly those who might not have had their work published previously, to submit proposals as we hope to be responsive to the needs of Environmental and Public Health practitioners. I am particularly keen that this series is seen as an opportunity for first-time authors and as ever would urge students (whether at first- or second-degree level) to consider this an avenue for publishing findings from their research. Why, for example, should the hard work that has gone into a dissertation or thesis lie unread on a library shelf? We can provide advice on turning a thesis into a book.

With the demise of the *Journal of Environmental Health* this Routledge series provides a route for practitioners to improve the profile of the profession. EHPs have not been good at telling others about their work and this failure may be one reason why they have not been deployed in any consistent way, nor their skills used most effectively. To be considered a genuine profession and to develop professionally, EHPs on the front line need to 'get published', writing up their work of protecting public health. This will allow them to analyse and report on what worked in practice, what was successful and what wasn't and why. This can provide useful insights for others working in the field and also highlight policy issues of relevance in Environmental Health.

Contributing to this series should not be seen merely as an exercise in gathering CPD hours but as a useful method of reflection and an aid to career development, something that anyone who considers themselves a professional should do. I am pleased to be working with Routledge to provide this opportunity for practitioners.

It is not intended that this series takes a wholly 'technical' approach but provides an opportunity to consider areas of practice in a different way, for example, looking at the social and political aspects of Environmental Health in addition to a more discursive approach on specialist areas.

Our hope remains that this is a dynamic series, providing a forum for new ideas and debate on current Environmental Health topics. If readers have any ideas for titles in the series please feel free to submit them to me as the series editor via the e-mail addresses provided below.

'Environmental Health' can be taken to mean different things in different countries around the world and so we welcome suggestions from a range of professions doing "Environmental Health" work or policy development. EHPs may be a key part of the Public Health workforce wherever they practise, but there are also many other practitioners working to safeguard public and environmental health. Thus, this series will enable a wider range of practitioners and others with a professional interest to access information and also to write about issues relevant to them. The format of the book means a relatively short production time, so contents will be more immediate than in a standard textbook or reference work.

Forthcoming monographs are likely to cover such areas as Environmental Health in South Africa, power-people-planet (on leadership in the climate emergency), and damp and health. We are also in contact with colleagues around the world, encouraging them to submit proposals. That does not mean we have no need for further suggestions, quite the contrary, so I hope readers with ideas for a monograph will get in touch via Ed.Needle@tandf.co.uk or Patrick.Hetherington@tandf.co.uk.

Stephen Battersby, MBE, PhD, FCIEH, FRSPH
Series Editor

Acknowledgements

We should like to place on record our very grateful thanks to all who broke off from their important work to record and share their 'COVID-19 Stories', particularly those that we badgered for further insights. There are too many to thank individually, but you know who you are and may well recognise your sentiments expressed in this book.

Contributors

Dr Rob Couch is an Environmental Health practitioner who has worked across the public, private, academic and charity sectors in the UK and overseas. This includes many health protection roles and therefore since February 2020 he has been part of the COVID-19 response for a Public Health team covering three local authorities in the East of England. Rob's PhD explored the street level work of local government EHPs in one of South Africa's largest and most unequal cities. His wider research interests include the Environmental Health workforce, the challenges of urbanisation and the climate emergency, particularly in African cities. Rob is also a co-founder of the UK Environmental Health Research Network that promotes the role of research, publication and a more evidence-based Environmental Health.

Dr Surindar Dhesi has a particular interest in Environmental Health and its role in public health, especially in tackling health and social inequalities and protecting the most vulnerable in society. She led the first Environmental Health assessment of the Calais refugee camp and is currently involved in projects investigating the impacts on well-being of living conditions experienced by displaced people in Indonesia, Nigeria and Gaza. Her doctoral research and associated Routledge monograph focussed on the role of Environmental Health in tackling health inequalities in England. Surindar is currently employed as a lecturer and course leader of the MSc Health, Safety and Environment Management at the University of Birmingham.

1 Countdown to the pandemic

An Environmental Health point of view

1.1 Introduction

Just over a year has passed since SARS-CoV-2 emerged and spread around the world as COVID-19 and is now doing so again through a second wave associated with new strains. In just 12 months it has brought many nations to their knees socially, economically and emotionally, taking away lives and livelihoods. In its wake, SARS-CoV-2 has put public health systems to the test like no other infectious agent since the influenza pandemic of 1918/1919. Some have buckled under the strain, while others have excelled, though there is no doubting the fact that none of this should have come as a surprise.

While our interest is in the professional response to COVID-19, in a sense it is the people's disease, as our media inter-connectedness caused the population at large to confront it. Whereas one might have been hard-pressed to discover many members of the public with more than a hazy idea of what 'Public Health' was, still less an awareness of who practised it and who spoke on its behalf, this is no longer the case. If the term 'epidemiology' might have evoked something mysterious and beyond the need for comprehension, today, its language punctuates everyday conversation, with the 'R-Value' much-quoted and the personalities attending press conferences identified by name, title and mannerisms.

However, in the time it has taken us to orbit the sun, COVID-19 has moved from being a condition of little real consequence to one whose devastating effects have been felt by billions. Together, as a world audience, we have watched while clinical science and biotechnology have played catch-up with a novel coronavirus in real time. Yet, all the while science has been hard at work devising effective treatment regimens and the 'magic bullet' of a vaccine, we have been reliant on the most basic of Public Health measures to keep us safe.

Among those responding to COVID-19 were a small but dedicated band of lesser-known health professionals that have been involved in

DOI: 10.1201/9781003157229-1

promoting and performing many of these fundamental measures so as to bring this disease under control. Their discipline is known as 'Environmental Health' (referred to throughout as 'EH') and they are called 'Environmental Health Practitioners' ('EHPs')

For EHPs, like so many others with a remit to protect health, and even more so those rescuing their fellows when they succumb to illness, this has been a testing time. Yet they have risen to the occasion, and despite finding themselves performing unfamiliar tasks, demonstrated their special blend of skills to critical acclaim. How EHPs were confronted by COVID-19 and emerged from the first wave as a force for good among the wider Public Health workforce is the subject of this book.

1.2 SARS-CoV-2 and the COVID-19 pandemic

The story of where this strain of coronavirus may have originated, and how the infection took hold and spread, prompting the stark realisation that the world was not immune, literally or metaphorically, to the direct and indirect effects of infectious disease, is a story already much told. We outline it here because it serves as a warning that, 'midst the excitement of discovering drug therapies that help to keep patients out of ICU and off ventilators, and vaccines that may prevent people getting sick in the first place, Environmental Health, and hygiene measures based on the principles of sanitary science, will remain the third and vital element of the solution for the foreseeable future.

While we are eternally grateful to medicine, and those that practice it, for keeping us going individually, medical science does not have all the answers when it comes to keeping populations safe and well. COVID-19 is not going to be the last infection associated with a zoonotic pathogen that threatens to break out from its point of origin and transmit serious disease worldwide, so we must better learn lessons from this one.

If we did not know it already, novel agents responsible for infectious agents such as SARS-CoV-2 require, by their nature, time to be identified and understood, and then more time still to develop the biomedical means to treat and ultimately prevent them. As we write, across the world, vaccines are being approved by national regulators, mass production is underway and supplies are being distributed and, in the vernacular of the moment, 'getting into peoples' arms'. Also, sadly, we are now seeing the outbreak of the World Vaccine War, as countries vie for their doses, with richer countries inevitably prevailing.

Clearly, its efficacy will only be properly known when a critical mass of the world's population has the protective antibodies, whether acquired naturally or through vaccination. Even then there will still be many unknowns, not least how long immunity will last and whether it is truly effective against onward transmission of mutant variants, though the signs are hopeful.

Given the sheer enormity of the problem and the response to it, we can only really touch on the key points, still less subject them to critical examination – the time for this will come. What we can do is set the scene for subsequent consideration as to how the Public Health service, and EH in particular, were called on to respond.

However, before we do so, some preliminaries are in order so that we can properly appreciate what EHPs seek to achieve under normal circumstances. Only then can one understand how practitioners of EH have gone about meeting the challenge of COVID-19. We apologise to those who know this already; however, recognising the distinction between different disciplines and areas of practice is vital to an appreciation of the response overall.

1.3 Public Health

In subsequent chapters we will be exploring how the COVID-19 epidemic has served to draw attention to different Public Health and Environmental Health interventions and brought practitioners of both disciplines, if they weren't already, closer together. To EHPs in countries where the two have always been practised as a single, integrated service, this may seem odd. However, for others, Environmental Health and Public Health were set on separate paths some years ago, though as we shall see, the pandemic has served as a reminder – if one were needed – that they come from the same stable.

Poignantly, 9 January 2020, the day a novel coronavirus was identified as the cause of a cluster of cases of pneumonia in China marked the centenary to the day of Charles-Edward A. Winslow's (1920) neat definition of Public Health as '… the science and art of preventing disease, prolonging life and promoting health through the organised efforts of Society' appeared in *Science*. However, to do justice to Winslow's vision for Public Health, we should consider that he went on to describe it as including '… the sanitation of the environment, the control of community infections, the education of the individual in principles of personal hygiene, the organisation of medical and nursing service for the early diagnosis and preventive treatment of disease …', thus capturing the characteristics of a discipline that has found itself coming to the fore as the COVID-19 pandemic gained traction.

Of course, the focus of disease management and control in the intervening 100 years has shifted markedly from the communicable to the non-communicable, and the effort that practitioners of Public Health invest in protecting health and well-being from diseases to do with genetic vulnerability, environmental exposure and the so-called diseases of lifestyle must not be overlooked in this present climate. In this respect they seek to bring about health improvement by promoting the value of good health through diet, exercise and so on, while drawing attention to inequalities and inequities, and advocating for the disadvantaged.

For now, our interest is in the control of respiratory disease, and so Winslow's observations remain as relevant now as they did when the world emerged from the 1918/1919 Influenza pandemic. Consequently, the simple rules of public health and hygiene he espoused, and which were put into action once the aetiology of the COVID-19 became known – 'test and trace', outbreak investigation, quarantine/isolation, raising hygiene awareness and so on – will need to continue indefinitely.

While the 'science' of Public Health is simple and straightforward to apply in theory, the 'art' is to know how to see these measures effectively and efficiently implemented. Anyone who has had experience of trying to impart information to someone who might have been convinced by a contrary argument, or to persuade them to exercise restraint or adopt a particular behaviour – especially if this means restricting their personal freedom, causing them financial loss and possibly additional hardship – will know it to be a skilled art.

1.4   Environmental Health

In seeking to represent the discipline we know as 'Environmental Health', and especially the activities of those that practise it, the 'new' definition arrived at by Ian McArthur and the late Xavier Bonnefoy in 1998 (based on the outcome of a WHO Consultation in 1994) is a good place to start. They saw it embracing, '… those aspects of human health, including quality of life, that are determined by physical, chemical, biological, social and psycho-social factors in the environment', which for practitioners of EH means applying '… the theory and practice of assessing, correcting and preventing those factors in the environment that can potentially affect adversely the health of present and future generations' (MacArthur & Bonnefoy, 1998).

We see from this the nature and extent of the challenge faced by practitioners of Environmental Health even without the complication of pandemic disease. This is especially the case for those working in wickedly poor and deprived areas of the world, where unequal access to the most fundamental elements of health – clean water and sanitation – compromises something as basic as handwashing as a means of interrupting viral transmission (Local Burden of Disease WaSH Collaborators, 2020).

For the purposes of this book, it is helpful to regard EH as a specialism of Public Health, where its 'science and art' lies in understanding those factors in the environment that can cause harm – sometimes referred to as the 'social determinants of health' – and then understanding the underlying reasons why this should be the case – sometimes described as recognising and seeking to resolve the 'causes of the causes of ill-health'. By doing so, and subsequently directing attention to those situations where the greatest need exists and greatest good can be affected, practitioners of the discipline also hope to go some way towards addressing inequities in health and redressing inequalities in health.

The COVID-19 pandemic has demonstrated the heightened risk of infection, serious illness and death are experienced by those already most vulnerable to suffering ill-health and premature mortality for other reasons and through other causes. Therefore, as the pandemic slowly wanes, it is beholden on practitioners of both Public Health and Environmental Health to bring their efforts to bear most decidedly on those likely to suffer from living in unsatisfactory housing, employed in cramped and poorly paid workplaces, on an income (if they have one) that deprives them and their families of a proper diet.

Nothing should ever be taken away from the curers and carers whose contribution to restoring our health must never be forgotten. However, the much used adage that 'prevention is better than cure' is as true now as it has ever been. Put simply, the question should not be whether we have enough ambulances at the foot of a cliff to take away those who keep falling off it, but to see why they are doing so in the first place, and then finding ways of preventing others from following suit.

1.5 The practice of Environmental Health

Traditionally based in local government and not a healthcare setting, Environmental Health practice draws on a strong association with the local community, and, where necessary to maximise impact, the development of effective community partnerships. As we shall see later, these close and long-standing practitioner-community links have proved vital in taking public health messages and instructions out to those who need to hear them most. As one practitioner from South Africa (SM35) put it to us in researching this book, having cause to deal with people in their homes, shops, churches and burial society meetings, no other health professional '… is able to reach so far and these were the main target areas …'. Yet in some parts of the world – here, thinking about England – it was quite clear from early on in the pandemic that EHPs with expertise in Public Health and '… getting messages out to the public' were not being deployed effectively' (Battersby, 2020).

'Environmental Health 2012: a key partner in delivering the public health agenda' (Burke et al., 2002), prepared on behalf of the Health Development Agency (HDA), explored as a positive 'Vision' the scope of Environmental Health services in the settings of the home, workplace and the living environments, where food, air, water and the land are challenged by the biological, chemical, physical, social and psycho-social stressors, with EHPs playing a central role in meeting these challenges.

While this book considers the roles performed by EHPs in coping with an acute respiratory infection, it should be remembered that their wider professional skills and expertise interests extend to the effects of exposure to multiple stressors, especially where issues of equality and social justice have been recognised as key aspects of public health for some time (Marmot, 2010), Their efforts should be considered alongside the role played by local authorities in public health which has continued to be championed for the importance of place-based approaches and community-based initiatives (Dombey & Bonner, 2020).

To this end, EHPs and their managers have had to decide how best to balance, and at times juggle, the need to remain attentive to the harms that might arise if they ignored their normal duties and responsibilities, and their instinct to become involved in the issue of the moment. That they have managed to do this, and act on their special skill of seeing a problem like COVID-19 in the whole, thus looking beyond the immediate interruption of its spread, and addressing issues that might interfere with the adoption of activities and behaviours that will see lives and livelihoods protected, is a demonstration of both the strength of the discipline and those that practise it.

1.6 Health protection

The vital health protection role – keeping people safe and well – especially when threatened by infectious disease, has been at the heart of Public Health since the efforts of Edwin Chadwick, Dr John Snow and Florence Nightingale sought to control infections in the mid-19th century. The story of how this function originally came into being, and, thereafter, developed to fit and suit the countries' needs and resources, is likely part of national folklore. Of course, not all countries discharge their Public Health and Environmental Health functions in the same way – populations rarely suffer the same degree of risk and vulnerability to infection even within the same country – so we can expect that COVID-19 will have more serious consequences for some countries than others.

Were this book about the changes made to the administration of public health in countries around the world over the past 20 years, and how these had impacted for good or bad on the response to COVID-19, we would have described systems that had become dissociated and their remits confused; where professional 'silos' and boundaries had built up; and where there had been a sustained lack of investment. It might also show an unwillingness to commit to forward planning and encouraging people and groups to reach a point of mutual understanding, and so trust, fundamental to the success of joint working.

1.7 The health protection 'vision' becomes blurred

Despite the hopeful 'Vision' expressed by the HDA in 2002, it was critical of many aspects of EH practice in the UK, observing that local authorities had become excessively focussed on statutory enforcement duties handed down by government. Since then, others have focussed on constraints imposed by diminished local government funding, the fragmentation of the public health professions, and issues around EH (in)visibility and recognition by decision-makers (Dhesi, 2018).

As we shall see, this sentiment was relevant globally and shared by many of those who contributed to this book, suggesting that EHPs had been passed over at the outset, or they had seen their efforts hindered, by what one respondent (SM26) thought was a lack of understanding on the part of senior managers of their role, though adding, '... neither does the Government'. However, many others confirmed that the pandemic has raised the profile of the profession, or as a local government manager in South Africa (SM30) put it, '... decision makers learned more about the value of the Environmental Health Service'.

The consequences of a failure to have in place the infrastructure to meet the challenge of pandemic disease in England, for instance, were raised since the current system of public health was established in 2013 with the creation of Public Health England (PHE). In evidence to the Communities and Local Government Committee in November 2012 (CLGC, 2013), Professor Gabriel Scally was of the view that the responsibility for dealing with local outbreaks should reside with those working locally, further adding,

> I am not convinced that if we have a problem across a substantial part or the whole of the country, Public Health England would be able to provide enough staff to lead in every local area. I believe it should be the local director of public health and the local authority that leads that function, but that is not clear to me at all.

Certainly these, and other, sage words, seem to have been heeded by Government, for, in September 2019, we see in PHE's 'Infectious Diseases Strategy 2020–2025', sub-headed 'Addressing urgent threats in the

21st century' (PHE, 2019), the need to recognise 'partners' with whom they should intend to collaborate. This refers by name to Directors of Public Health, the Local Government Association, local resilience fo-rums (LRFs) and local health resilience partnerships (LHRPs), mention-ing, also, 'local authorities' and 'other public health agencies', though making no reference to Environmental Health or Environmental Health Practitioners.

Subsequently, in the 'UK Coronavirus Action Plan: A guide to what you can expect across the UK', which was made available on 3 March 2020 as a Policy Paper issued by the DHSC (2020a), paragraph 3.2 provided an assurance of the Government's preparedness (albeit for an influenza pandemic) and that their plans had been '… regularly tested and updated locally and nationally to ensure they are fit for purpose'. However, while we were told that LRFs and LHRPs (and NHS emer-gency planning structures in Wales) would engage with 'local organisa-tions' in taking '… primary responsibility for planning and responding to any major emergency, including a pandemic', it makes no reference to local authorities, Directors of Public Health by name, or Environmental Health. This would suggest that the intention in the first week of March was for the Government to entrust the role in England to PHE working in conjunction with the regional resilience fora/partnerships.

If the views of the many who have criticised the handling of the pandemic in the UK and elsewhere are heard, then this oversight will surely be made good, since Environmental Health and EHPs have clearly played an integral part in the public health response to COVID-19. This should not be forgotten as countries around the world revise their plans for the management of epidemic infectious disease.

1.8 Overview of the response to COVID-19

Many 'timelines' are available that map the emergence of the infection from its supposed source, some plotting the response to this by govern-ments around the world. We would not presume to repeat the exercise, referring anyone that wishes to learn more about the chronology in the first nine months through to mid-November 2020 from the World Health Organization's 'Timeline: WHO's COVID-19 response' (WHO, 2020a).

There is an argument for starting any 'timeline' on 30 January 2020, when the director-general of the WHO declared the novel coronavi-rus outbreak a public health emergency of international concern – the WHO's highest level. However, any chronology should really begin when the WHO's country office in the People's Republic of China picked up a medical report of 'viral pneumonia' in Wuhan, Hubei Province, which it passed on to the Western Pacific regional office, thus 31 December 2019.

1.8.1 January

In the perfect world, intelligence of this sort should have been communicated and picked up by every country on New Year's Day. However, if it was missed then, it should have happened by the third week of January when there was every indication of a potentially fatal respiratory disease, with certain evidence of onward human-to-human transmission in China. A Comment piece published online in *The Lancet* on 24 January 2020 should have been the 'wake-up call' to all outside China, given its title 'A novel coronavirus outbreak of global health concern' (Wang et al., 2020).

Reflecting on lessons learnt from past coronavirus epidemics, they wrote,

> The international spread of SARS-Cov in 2003 was attributed to its strong transmission ability under specific circumstances and the insufficient preparedness and implementation of infection control practices.
>
> Considering that substantial numbers of patients with SARS and MERS were infected in health-care settings, precautions need to be taken to prevent nosocomial spread of the virus.

And warning against inaction and delay they commented,

> We need to be wary of the current outbreak turning into a sustained epidemic or even a pandemic.
>
> We have to be aware of the challenge and concerns brought by 2019-nCoV to our community. Every effort should be given to understand and control the disease, and the time to act is now.

1.8.2 February

February saw the WHO making repeated attempts to draw attention to the seriousness of the matter and urging countries to make plans, raise their level of preparedness and put in place robust public health measures. There was clear emphasis on the need for decisive leadership, thoroughness in public health work and societal engagement (WHO, 2020b). This was especially the focus in the WHO document 'Operational planning guideline to support country preparedness and response' issued on 14 February 2020, where under Pillar 3 (of 7) entitled 'Surveillance, rapid response teams, and case investigation', the WHO warned '… countries with high-risk of imported cases or local transmission' to focus their surveillance on 'rapid detection of imported cases, comprehensive and rapid contact tracing, and case identification', essential '… to calibrate appropriate and proportionate public health measures' (WHO, 2020c).

1.8.3 March

March will likely be considered the most momentous weeks in the history of Public Health for a century. In countries around the world there was a realisation that there could be no further prevarication; what they chose to do, or not do, now might have terrible consequences – human and economic. In the UK, the message on 5 March 2020 (BBC, 2020) was that, according to the chief medical officer, the move from the 'Contain' (suppress) to the 'Delay' (mitigate) phase was 'highly likely', a sentiment with which we were told the Government broadly agreed, but which the prime minister still believed could be achieved without taking Draconian action that would have a damaging effect on the economy, or as he put it, 'striking a balance'.

Yet, all the while countries were toying with the idea, the World Health Organization, in declaring COVID-19 a 'pandemic' on 11 March 2020, was urging them to remain faithful to the WHO mantra of 'find, test, trace and isolate'. In his opening remarks to the media briefing that day, director-general said,

> We cannot say this loudly enough, or clearly enough, or often enough: all countries can still change the course of this pandemic. If countries detect, test, treat, isolate, trace, and mobilize their people in the response, those with a handful of cases can prevent those cases becoming clusters, and those clusters becoming community transmission. Even those countries with community transmission or large clusters can turn the tide on this virus.
>
> WHO (2020d)

The announcement of the decision to move to the 'Delay' phase came on 12 March 2020 in the UK with a press release from the Department of Health and Social Care (DHSC, 2020b) indicating that the decision had been taken '… in response to the ongoing coronavirus (COVID-19) outbreak'. Nevertheless, the UK chief medical officer warned against doing this too soon on the grounds that it would have a '... huge social impact' and that they needed to '... continue to do the right thing at the right time', but this would be '... based on careful modelling', and the measures themselves '... supported by clinical and scientific evidence'.

That 'careful modelling' came on 16 March 2020 with the 'bombshell' paper from the modellers at Imperial College London led by Professor Neil Ferguson entitled 'Impact of non-pharmaceutical interventions (NPIs) to reduce COVID-19 mortality and healthcare demand'. This offered governments around the world a stark choice – 'mitigate' (slow down the spread) of the infection, reduce the demand on critical care beds by two-thirds and halve the projected mortality if left uncontrolled, or seek to 'suppress', that is, to seek to reverse the growth of the epidemic and eventually eliminate the infection. However, concluding that mitigation in the UK would still leave over 250,000 dead, 'suppression' was presented as the only sensible option '... minimally requiring a combination of social distancing of the entire population, home isolation of cases and household quarantine of their family members' (Ferguson et al., 2020).

1.9 The book in outline

Looking ahead, Chapter 2 explains why we are focussing on the work of Environmental Health practitioners, the methods used to create the book and how we managed the inevitable differences in EH practice across the world. In Chapter 3 the focus moves to critique the preparedness of Environmental Health for the pandemic, how COVID-19 prompted a re-think of Environmental Health priorities and led to new ways of working, before summarising four regional perspectives on the initial response. Chapter 4 outlines the response of EHPs to COVID-19 across their main areas of work, before focussing on specific health protection interventions and the wider roles of Environmental Health academics and the military response.

In Chapter 5 we summarise the challenges and opportunities experienced by EHPs, before questioning whether the pandemic might have had a favourable professional outcome for EH. Chapter 6 explores the role assumed by national and international organisations in supporting and representing EHPs, and how some UK EHPs had created their own support and advocacy networks. Chapter 7 discusses the preceding chapters, consolidating and explaining the main themes that characterise the global response so far, while the final chapter, Chapter 8, brings these together in a series of conclusions, before making recommendations for consideration as we hopefully put SARS-CoV-2 behind us.

Throughout, we make the point that this book is about a short period of what is now history. Though we believe it provides a valuable insight into the role performed by EH and EHPs, it never loses sight of the fact there is still much to share and learn from the global Environmental Health community.

References

Battersby, S.A. (2020). *Letter to The Guardian Newspaper* on 3 April 2020. Available at: https://www.theguardian.com/world/2020/apr/03/coronavirus-a-test-strategy-that-failed

BBC News (2020). *Coronavirus: UK moving towards 'delay' phase of virus plan as cases hit 115* (5th March 2021). Available at: https://www.bbc.co.uk/news/uk-51749352 (accessed: 8th February 2021)

Burke, S., Gray, I., Paterson, K., & Meyrick, J. (2002). *Environmental Health 2012: A key partner in delivering the public health agenda.* Health Development Agency. Available at: file:///C:/Users/user/AppData/Local/Temp/environmental_health_2012.pdf (accessed: 9 February 2021)

Communities and Local Government Committee (2013). Session 2012/13 – Eighth report: *The role of local authorities in health issues* – Based on oral evidence given by Prof. Gabriel Scally on 29th November 2012 and detailed in www.parliament.uk – Minutes of Evidence HC694. Available at: https://publications.parliament.uk/pa/cm201213/cmselect/cmcomloc/694/121126.htm (accessed: 10 February 2021)

Department of Health & Social Care (2020a). Policy paper – *UK coronavirus action plan: A guide to what you can expect across the UK.* Available via GOV.UK at: https://www.gov.uk/government/publications/coronavirus-action-plan

Department of Health & Social Care (2020b). *COVID-19: Government announces moving out of contain phase and into delay,* Press Release: 12 March 2020. Accessed via GOV.UK and available at: https://www.gov.uk/government/news/covid-19-government-announces-moving-out-of-contain-phase-and-into-delay (accessed: 9 February 2021)

Dhesi, S. (2018). *Tackling Health Inequalities: Reinventing the Role of Environmental Health.* Abingdon: Routledge.

Dombey, R., & Bonner, A. (2020). A place-based approach to healthy, happy, lives. In A. Bonner (Ed.), *Local Authorities and the Social Determinants of Health,* 105–120. Bristol: Policy Press.

Ferguson, N.M., Laydon, D., Nedjati-Gilani G., et al. *Impact of non-pharmaceutical interventions (NPIs) to reduce COVID-19 mortality and healthcare demand.* Imperial College London (16th March 2020), https://doi.org/10.25561/77482. Available at: https://www.imperial.ac.uk/media/imperial-college/medicine/mrc-gida/2020-03-16-COVID19-Report-9.pdf (accessed: 10 February 2021)

Local Burden of Disease WaSH Collaborators (2020). Mapping geographical inequalities in access to drinking water and sanitation facilities in low-income and middle-income countries, 2000–17. *The Lancet Global Health,* 8(9), e1162–e1185. https://doi.org/10.1016/S2214-109X(20)30278-3, Open Access, September 2020. Available at: https://www.thelancet.com/action/showPdf?pii=S2214-109X%2820%2930278-3 (accessed: 9 February 2021)

MacArthur, I., & Bonnefoy X. (1998). *The Concepts of Environmental Health, Environmental Health Services in Europe 2: Policy Options.* Copenhagen: World Health Organization Regional Publications, European Series No.77. Available at: https://www.euro.who.int/__data/assets/pdf_file/0003/156927/e58195.pdf (accessed: 9 February 2021)

Marmot, M. (2010), *Fair society, healthy lives: Strategic review of health inequalities in England post-2010.* UCL. Available at: http://www.instituteofhealthequity.org/resources-reports/fair-society-healthy-lives-the-marmot-review/fair-society-healthy-lives-full-report-pdf.pdf (accessed: 9 February 2021)

Public Health England (2019). *Infectious diseases strategy 2020–2025: Addressing urgent threats in the 21st century.* Available at: https://assets.publishing.service.gov.uk/government/uploads/system/uploads/attachment_data/file/831439/PHE_Infectious_Diseases_Strategy_2020-2025.pdf (accessed: 1 December 2020)

Wang, C., Horby, P.W., Hayden, F.G., & Gao, G.F. (2020). *A novel coronavirus outbreak of global health concern* – Published online: 24 January 2020, https://doi.org/10.1016/S0140-6736(20)30185-9 (corrected: 29 January 2020) and in print – Comment: *The Lancet,* 395(10223): 470–473, 15th February 2020. Available at: https://www.thelancet.com/action/showPdf?pii=S0140-6736%2820%2930185-9 (accessed: 10 February 2021)

Winslow, C-E.A. (1920). Untilled fields of public health. *Science,* 51(1306), 23–33. https://doi.org/10.1126/science.51.1306.23

World Health Organization (2020a). *Timeline: WHO's COVID-19 response.* Available at: https://www.who.int/emergencies/diseases/novel-coronavirus-2019/interactive-timeline#! (accessed: 10 February 2021)

World Health Organization (2020b). *Listings of WHO's response to COVID-19.* Available at: https://www.who.int/news/item/29-06-2020-covidtimeline (accessed: 24 February 2021)

World Health Organization (2020c). *Operational planning guidelines to support country preparedness and response* (14 February 2020), WHO, Geneva. Available at: https://www.who.int/docs/default-source/coronaviruse/covid-19-sprp-unct-guidelines.pdf?sfvrsn=81ff43d8_4 (accessed: 10 February 2021)

World Health Organization (2020d). *WHO Director-General's opening remarks at the media briefing on COVID-19* – 11 March 2020. Available at: https://www.who.int/director-general/speeches/detail/who-director-general-s-opening-remarks-at-the-media-briefing-on-covid-19---11-march-2020 (accessed: 20 February 2021)

2 Discovering an Environmental Health perspective on COVID-19

2.1 Focus and background of the book

This book focusses on how the Environmental Health services in different countries have been mobilised – in particular, how practitioners have been deployed and responded to the challenge. By drawing attention to the work performed by EHPs during the first nine months of the pandemic, the intention was to provide direction and inspiration to the profession for how it might, working alongside other public health professionals, tackle any resurgence of pandemic SARS-CoV-2, and whatever might follow. It would also be an invitation to governments, municipalities, voluntary organisations, and businesses to make better use of environmental health practitioners as front-line guardians of the public's health.

Although some comparisons between different countries are inevitable (and, in some cases, between different federal states and regions in one country) no direct judgement is made on a state's overall response to the pandemic. COVID health data may measure more than a country's performance in infection control, overlooking the fact that the virus may have entered some countries insidiously, and others at scale. Nevertheless, several of our respondents drew attention to the spread of infection in the early stages of the epidemic in their countries which they believed pointed to the level of preparedness to initiate effective and timely interventions, and these couldn't be ignored.

The authors are British but are international by inclination. There may be more written about the UK's response as the authors have more direct experience on this, but the UK itself is made up of four countries with devolved powers to assemblies or parliaments in Scotland, Wales, and Northern Ireland (there are also Crown dependencies of the Isle of Man and the Channel Islands) which makes it difficult to speak of the experience of EHPs across the UK. Moreover, there are 193 member states of the United Nations, and, consequently, every attempt has been made to avoid making the text too Anglo-centric, since the activities

DOI: 10.1201/9781003157229-2

of EHPs outside the UK are every bit as important and valued. As one peer reviewer of the proposal said about the book in plan, this is '... a great opportunity to publish a text that offers a particular snapshot of this moment in history', focussing, as another put it, through '... a truly global lens'.

2.2 Targeting 'practitioners of Environmental Health'

As the book is intended to present as wide a range of experiences as possible, careful thought was given to the different job titles under which practitioners are known. Just in one country, indeed in neighbouring local authorities, those performing roughly the same job can be known by quite different titles. Conversely, a similar title might disguise the obvious fact that the needs and demands of the populations served by EH may be markedly different in different places. Thus, for instance, practitioners working in west Africa and the townships of South Africa face far more fundamental public health challenges than those in western Europe and southern England, and so the complications arising from COVID-19 might be expected to be different too.

Consequently, the authors sought to dispel any pre-conceived notion of what might represent 'typical' Environmental Health practice and there being a 'typical' EHP. We also felt it unnecessary to make a distinction between the work performed by EHPs in the private and the public sectors, as some EHPs are employed in both. What was required were first-hand accounts from those who, though they might not be called Environmental Health Practitioners, were 'practitioners of Environmental Health'.

Thus, 'Environmental Health Practitioner' is used in a generic sense, indicating in our invitations to contribute to the book that it was what they did that mattered, not what they were called. In addition, the experiences told to us first-hand by practitioners in the field are combined with any that had been responsibly acquired and properly documented elsewhere if they addressed the matters set out in our invitational letter.

Fears that such differences in the experience of EHPs worldwide would be such that it would be impossible to bring them together proved unfounded. In fact, the marked similarity in the experiences of practitioners in different countries and settings was the revelation, and they were the more significant because of the differences in the work these practitioners perform under normal circumstances.

The experience of the authors seeking to secure material for this book, as with other works, is that EHPs are reluctant to describe what they do and how they see it contributing to health, especially if it might be seen as 'courting praise'. Some recognise this trait in themselves and colleagues, going so far as to say that this might be part of the reason why they are passed over at times and the profession fails to receive the credit it deserves. Thus, when we received descriptions of personal achievement, and the sense of pride and satisfaction in a job well done, we considered it less likely to be exaggerated or self-congratulatory.

2.3 Informing the book

From the outset the authors wanted to let EHPs working wherever and in whatever capacity to tell their own story in any way they felt comfortable. Efforts to engage with them and seek information on their experiences were made through invitations posted in online and print journals, including the CIEH's *Environmental Health News*, through the International Federation of Environmental Health's membership network, social media platforms (LinkedIn and Twitter) and by approaching individuals directly by e-mail.

It was accepted from the outset that it would be a challenge seeking to gather information from practising EHPs while the pandemic was still in progress, and this became more problematic as countries worldwide were beginning to experience a resurgence in cases: the 'second wave' of the disease. However, rather than acting as a disincentive, it served to encourage us to capture the experience of practitioners during this opening phase of the pandemic after which their accounts might be confused by subsequent events. That this might help shape future policy and practice, if and when this situation arises again, was an important consideration. Key to this was not simply enquiring what they had been doing, but to invite them to share their thoughts and feelings about the obstacles faced, outcomes achieved (or not) and frustrations felt.

If this phase of information gathering that began in late September 2020 produced relatively few responses, they provided some especially important, candid, and revealing insights. Only in a few cases were respondents instructed by their managers to withdraw material submitted because of the sensitivity of the subject under discussion. However, the clear majority were content with the assurance that identities would be withheld unless they expressly agreed to waive anonymity, or their comments were rendered unidentifiable.

In addition to the responses received from EHPs and Public Health practitioners through this means, a short SurveyMonkey questionnaire was released in early November, which asked the recipient to identify where and in what capacity they were working; explain the nature of the COVID-work with which they had been engaged from the outset; outline

Table 2.1 Summary of the countries and sectors of the 62 EHPs from 12 countries/regions that responded to the survey

Country	Sector (number)							
	Local government	Regional government	National or regional government agency	Private	Healthcare	NGO	University	Military
Australia	1	1					1	
Germany				1				
Liberia			1					
The Netherlands		1	1					
Nigeria			1					
Portugal	1						1	
Slovenia							1	
South Africa	9		1		1		3	
UK	14	1	3	5			2	
(Channel Islands)			3					1
The United States	2		2			2		
Zambia				1				
Not stated	1							
Total	28	3	12	7	1	2	8	1

the impact this had (and was having) on their non-COVID workload; assess the influence of their past environmental health education, training and experience on their capacity to perform COVID-related activities; and describe the greatest challenges and opportunities they had encountered thus far. It concluded by asking whether they believed that the contribution EH had made in responding to the COVID-19 pandemic would serve to strengthen the position of the Environmental Health profession in their country.

Conscious of the pressure on their time, the survey was designed to be completed in around 10 minutes, before being released through the means used previously, but with the addition of e-mails sent directly to membership organisations in the IFEH network and Facebook (Table 2.1).

Finally, we drew on magazine and journal articles, research papers and updates compiled by national bodies that commented how public health in general, and EH/EHPs in particular, had responded to the COVID-19 pandemic.

2.4 Reconciling global differences in Environmental Health practice

Even if there is a common understanding of the principles of Environmental Health, different countries around the world have rather different ways of delivering services based on these principles. They use legislative powers differently and entrust responsibility for discharging these through mechanisms and personnel peculiar to each country. Consequently, we did not set out to deliberately tease these out, service by service, and country by country, though some of the material received made these differences apparent. Our focus was fixed throughout on the circumstances that EHPs found themselves in these early weeks and months of the pandemic both in terms of their normal health protection role and in the context of COVID-19. By exploring how they had become involved in helping to control onward transmission of the virus and thus prevent illness, it was hoped that we might establish the Environmental Health perspective on the pandemic, and its practitioners as a global force in health protection.

Responses have been collected from EHPs working under most forms of administrative set-up ranging from those where duties and responsibilities are delivered through a devolved civil service directed from the centre, through elected local government underpinned by national legislation, to autonomous federal systems of independent states working to an amalgam of central and local edicts. Our research is a reminder of how different these are, and therefore all the more remarkable that a coherent picture emerged.

2.5 Organising the responses

By providing respondents with an opportunity to either express themselves in free text or respond to 'open-ended' questions, it was anticipated that this might produce data in a form that would be difficult to unravel and develop around common 'themes'. However, in very few cases was it not possible to translate the original responses into more concise narratives. Once assembled, these offered an insight into the work performed by Environmental Health departments, divisions and the practitioners working within them in a wide range of settings and capacities around the world. A simple coding system was applied to the responses/respondents, and these were held (and will remain held) securely by the editor.

If lessons are to be learned from experiences of EHPs, it would be wrong to exclude any contributions which make constructive criticism of governments and institutions for their policies and actions. That said, we felt it inappropriate to admit a 'rant' that vilified any administration or individual for simply making an ill-advised decision. We had to remain focussed firmly on the personal reflections of practitioners on their COVID-19 experience – what they did, what they thought they had achieved, and how their contribution might have made a difference.

2.6 Post-script

At the time of finalising the text for this book (January 2021), in Europe, the second wave is at a peak, fuelled by new variants which have greater transmissibility and may be able to evade vaccine-induced antibody responses. In September 2020, at the outset of drafting the manuscript, the second wave was just emerging and there could not have been a worse time to ask practitioners to break off from their work long enough to reflect and ruminate. Yet they did, and we are indebted to them for doing so.

3 Early impact of COVID-19 on Environmental Health practice

3.1 Introduction

As the book title suggests, we set out to represent the experiences of EHPs employed in various guises and in different settings, describing what they did before COVID-19; how their work was initially transformed by COVID-19; and how their qualifications and experience in EH served them to make this transformation. It soon became clear from the accounts that the pandemic had established new ways of thinking and doing, and these are introduced here. In addition, we had the benefit of accounts prepared by, or on behalf of, EHPs, that served to amplify the personal stories received.

In the next chapter we look in more detail at the sorts of activities EHPs found themselves performing, and how being so engaged in COVID-related work had affected the discharge of their normal duties and responsibilities. Then, in Chapter 5, we will look at the challenges and obstacles they encountered, how the pandemic provided them with opportunities to demonstrate their capabilities, and how they saw these serving to advance the cause of EH.

What is immediately clear from the accounts received is that in many cases, though not all, their work has become more varied, wider-ranging and far more challenging. It included what might be described as their 'normal work', supplemented, to a greater or lesser extent, by 'COVID duties'. These varied between enforcing restrictions on commercial activity alongside their regular duties, to work expressly focussed on controlling onward transmission of the virus.

Of course, how EH responded was likely to reflect the rate that coronavirus became established in their country or region, and its growth in prevalence, and also an assessment of the consequences of abandoning their 'normal' work in, for instance, housing, environmental protection and food safety. This should be borne in mind as we consider how EHPs responded to the unfolding crisis.

DOI: 10.1201/9781003157229-3

Not surprisingly, Public Health practitioners in countries affected sooner by the infection tended to be mobilised sooner. However, with a few notable exceptions, there is little evidence from our EHP respondents that they were actively involved in COVID-19-related work much before March, save for some who said they saw the pandemic coming and pressed their Public Health authorities and local councils to start planning a co-ordinated response.

3.2 Work undertaken before the COVID-19 epidemic took hold

Of the work performed by our respondents prior to the arrival of SARS-CoV-2 in their respective countries, we managed to capture responses from those engaged in most of the roles expected to be performed by EHPs in the public sector. This ranged from those field-based practitioners involved in both 'general practice' to senior executives involved in strategic planning and policy development. In addition, we received responses from academics based in universities teaching and researching environmental health, and from the military cadre.

With a few exceptions the international 'language' of environmental health made our understanding of their role, and the nature and purpose of their work, reasonably straightforward to represent on their behalf. If there had been more time one would have wanted to explore in greater detail the challenges they faced before, and then once the COVID-19 pandemic took a hold. However, for the purposes of this book it was enough to accept that if they described that they were now only performing, say, 10% of their routine work, then what they were now doing was in very marked contrast to this.

3.3 Preparedness to respond to the COVID-19 pandemic, and why?

Given the opportunity to assess one's own capability and competence to respond to an emergency or unfamiliar situation, it would be unlikely that an individual, still less a health professional, would adjudge themselves incapable or incompetent. Yet, with the absolute assurance of anonymity, and the opportunity to describe their preparedness in terms of their own background education, practical training and professional experience, we obtained an outcome more revealing than we had expected.

Our respondents described exercising a wide-ranging set of skills and having to draw from knowledge that they may not have had cause or reason to use for many years, perhaps as far back as their initial training. By some margin rather more EHPs considered themselves well prepared to meet the challenges presented by the pandemic than not, one express-ing surprise, saying 'Better than I could have expected' (SM54 – South Africa), while others used superlative terms such as 'massively' when re-ferring, in particular, to contact tracing (SM8 – UK), 'hugely', in helping to start a 'track and trace' system in the Channel Islands (SM6), and variously 'excellently' (SM15 – United States; SM36 – South Africa; SM56 – UK) and causing them to be 'really/very well' prepared (SM4 and SM57 – UK; SM50 – Australia, SM53 – South Africa), and that their EH qualification, training and experience had proven 'essential' (SM64-UK), 'invaluable' (SM63 – UK) and 'extremely helpful' (SM46 – UK), with a West African EHO (SM16) concluding that their background in environ-mental health had prepared them '…very well for the great task ahead'.

Respondents seemed to welcome the opportunity to develop this further by describing why they considered this important, setting out the specific areas of practice or technical competence where this had been demonstrated. In this respect they regarded their 'understanding of risk and control' (SM57 – UK), ability to know how and when '… to elevate person-to-person contact during the early community transmission phase of the pandemic' (SM50 – Australia) and to conduct outbreak control work (SM4 – UK) as being valuable, along with the ability to read and understand legislation, and distinguish this from guidance (SM62 – UK).

On a personal level, respondents in the UK were especially aware of their 'transferable skills' (SM8 and SM27); the capacity to swiftly grasp the extent and complexity of a situation by applying a '… holistic approach' to problems and problem-solving (SM59); using their 'interview and investi-gative skills' (SM41), being able to 'work under pressure' (SM57), and able to 'think outside of the box' (SM55), as significant qualities.

A principal health protection practitioner with Public Health England (SM23), though formerly an EHP in local government practice, consid-ered this to be responsible for their 'agility' – a necessary attribute given the ever-changing nature of service delivery; 'pragmatism' – necessary in distinguishing between 'urgent' and 'important' tasks; and 'results fo-cussed', in this case '… concentrating on actions that protect health'.

However, in common with those EHPs who contacted the authors directly, survey respondents found their EH qualification, and the skills acquired through their regulatory and advisory work, were important. This, together with their local knowledge of their communities and business sectors in the area and an understanding of infectious disease control (particularly the impact of human behaviour), had put them in especially good stead.

How some of these 'hard skills' were put to good use is explored later, but respondents seemed keen to expand on how their 'soft skills' had served them just as well. So, as a UK EHP (SM62) put it, their ability to 'engage with business and the public' and skills, among others, acquired '... in training and as an EHO in practice' were valuable assets, as were 'people skills' which had proven '... invaluable to provide reassurance and encouragement at such a time of anxiety and uncertainty'.

One senior EHP in the UK (SM42) who considered their 'communication skills' the most useful tool observed that colleagues from other departments drafted in to undertake community resilience work simply did not have the skills to deal with '... the myriad of issues that talking to elderly and vulnerable people brings', going on to identify these as '... not wanting to admit that they need help ... trying to take advantage, forgetfulness, loneliness and the sheer volume of inconsequential chat'.

3.4 Reasons for lack of preparedness

While most respondents felt personally ready and able, some were not, feeling this was due to shortcomings in their environmental heath qualifications and/or a lack of experience since qualifying. One, a specialist in emergency planning working at senior executive level (SM47 –United States), considered the baseline qualification fell short of preparing EHPs well for undertaking COVID-19-related work which they felt required '... infection control knowledge beyond EH training', leading them to suggest that '... programs should consider adding more content in the infection control area'.

Others, while not so critical of the baseline qualification and professional training of EHPs, suggested the need for post-graduate study as their practical training since their EH education did not fully prepare them, with SM39 in the United States crediting their Master of Public Health degree for providing a '... strong background in epidemiology'.

When a university in Australia ran a workshop on what EHOs would like to see included in courses after COVID-19 the delegates identified infection control and '... courses in mental health issues, dealing with aggressive people etc. [and] media communication' (PC12). However, a specialist EHP working with hospitality businesses (SM60 – United States) said that they felt that their preparation to perform the activities demanded from the pandemic came from their '... many life experiences' including working in a call centre and undertaking community support work after studying for a degree related to social care.

3.5 Reasons for heightened preparedness

Not surprisingly, the greatest preparation for the COVID-19 pandemic considered by those who had obtained it was previous experience in dealing with pathogens of comparable infectivity and seriousness, especially in emergency situations. In some cases, it was the fact that they had chosen to take additional qualifications, and in the case of SM5 in the UK, the experience of 34 years in local government, had prepared them 'very well' for the pandemic, so providing them with an '… opportunity to shine professionally'.

A specialist normally working in food safety and vector control cited work related to animal and human rabies, plague and other zoonotic diseases making it possible to move readily across into COVID-19 surveillance and control (SM38 – United States), and another, trained in Public Health, who spent time working in TB and malaria control in a neighbouring country, was able to take on a very significant COVID-19 project back in February which served to promote '… the value of Environmental Health' in the minds of decision-makers (SM30 – South Africa).

For an Australian specialist (SM40) it was working in logistics with Médecins Sans Frontières in West Africa during the Ebola virus epidemics that served them well when the pandemic arrived in Australasia. Similarly, a West African government official, with immediate experience of responding to Ebola, felt amply prepared and ready to do so, though thinking that the Government was 'not up to the task', and unprepared to renumerate in a timely manner those that were '… trained to do the job' (SM31)

For a South African EHP (SM35), the significant issue when it came to COVID-19 was one of scale, and despite their baseline qualification and experience serving them well in dealing with local outbreaks of food-borne illness and cholera, uncertainty over mode of transmission and the rapidity of spread meant that this was a situation for which no one could be prepared. This conclusion was shared by a practitioner from Portugal (SM33), who, despite undertaking an internship in Public Health and undertaking investigations into outbreaks of Legionnaires' disease, concluded that this was '… something so unexpected and … different from everything so far' that they we reliant on learning something new every day.

A rather different type of experience was described by the director of a consultancy operating internationally (SM37) when they, having expressed the belief that their environmental health degree had provided 'very little' of value in the present situation, asserted that 20 years of working with the hospitality industry globally and undertaking '… a wide range [of] outbreak investigations …' had secured them a '… broad understanding of epidemiology'. Their experience of dealing with outbreaks over the years had, they said, given them '… an insight into what controls should be recommended and implemented prior to governmental guidance being issued'.

3.6   The initial impact of the pandemic on routine (non-COVID) work

When EHPs found themselves confronted by COVID-19, they and their managers were faced with some difficult decisions to make regarding the discharge of their normal duties and responsibilities – either to remain broadly focussed on minimising the harm caused by the risk associated with their regular work or shift their attention predominantly, if not exclusively, to preventing infection and onward transmission of the virus.

The complex and rapidly developing situation presented by coronavirus clearly came as a shock to most local government authorities and their managers, prompting some to discontinue anything that might prove a distraction, or else promote the 'urgent' over the 'important', and devote their time and energy to exploring options to mitigate unfamiliar situations and problems by unconventional means. Others appear to have adopted a more conservative approach and looked for a way of continuing to meet their normal objectives and outcomes by adapting their service plan, 'scaling back' on their routine work (SM59 – Channel Islands) and then devoting all remaining time and resources to COVID activities.

No evidence emerged from the study to suggest that local authority EH managers or their staff chose to 'sit on their hands' in the hope that COVID-19 would simply go away; though there was criticism subsequently of the response of some local authorities when approached to release staff. For some, there was clearly a sense of frustration that local authorities were not reading the situation with the same degree of seriousness as those with a public health remit. In the words of SM63, an environmental health manager in the UK: 'In January, I was trying to get my council to start planning for a pandemic …'.

This migration from routine work to 'COVID work' seems to have taken place to a variable degree and at an uncertain pace, but several respondents sought to put a figure on how this impacted on what SM56 – UK) described as 'non-pandemic work' which had reduced to 'next to nothing', perhaps '5–10%', '10%' (SM51 – UK), others going further by saying that they were very soon fully committed to COVID work, all routine work having been discontinued (SM35 – South Africa, SM41 – UK, SM52 – not stated).

The consequential backlog of work caused some practitioners and their managers to seek to resume their statutory or regulatory duties as soon as it was considered reasonable to do so. However, it seemed that well into November 2020, EHPs that had migrated into COVID work in the spring were still so involved after the first wave had passed. A small number reported that there had been an expectation early on, and in some cases throughout the first wave of the pandemic, that departments, and by extension, the officers, would be able to continue delivering their normal service and COVID-19 work.

At the time they responded to the invitation to share their experiences in November, one Team Leader (SM55-UK) described themselves as '… the lead person in the council for COVID as well as continuing to deliver the rest of the service', which included port health. Not surprisingly, the same respondent considered their greatest challenge was 'the very heavy workload'. The notion that departments might be able to continue in 'business as usual' mode was gradually eroded as time went by, though in some cases there was an acceptance that 'things may take longer than normal and we would not be providing any visits unless it was an emergency' (SM42-UK).

On the other hand, some EHPs found themselves being rapidly redeployed to perform duties quite different from their 'normal work', such as SM39 – United States, who found themself setting up, and thereafter expanding to scale, a testing facility, and SM47, also in the United States, recruited to undertake COVID-related projects, some externally funded, by, for instance, CDC. Naturally, these opportunities tended to fall to those with a background in a related field, or else experienced at working at a more strategic or policy level. So, understandably, a senior manager normally responsible for directing waste management operations at provincial level in South Africa (SM36) was as soon as February '… co-ordinating … and guiding municipalities regarding [their] waste management response' to COVID-19.

A private sector Public Health specialist in Germany (SM28), employed in the cruise industry reported that, from February 2020, they have been involved in managing ships that were no longer in guest operation, implementing appropriate public health controls on ships in 'operational pause', conducting on-going health surveillance and outbreak investigation of the ships' company, engineering and designing '… "return to guest service" operational procedures, creating necessary training and health promotion material for crew members [and] conducting audits onboard or remotely to assess compliance'.

If there was a feature that characterised both the nature and rate of the response to the emergent pandemic it was that those EHPs working in less formal local government systems seemed to have felt able to exercise greater discretion, thereby acting and reacting more decisively when it came to abandoning 'business as usual' and responding to the immediate challenge of COVID-19. Some, bound by a hierarchical management structure and likely answerable to elected members, and therefore expected to carry as normal unless instructed otherwise, found this a frustration. One, an Australian specialist logistician SM40), found the bureaucracy associated with the need for the whole state government to respond as one hampered their ability '... to pivot quickly'.

For a former local government EHP, latterly working as an occupational health and food safety consultant (PC1), his 'COVID story' had been eventful following the discontinuation of his face-to-face work. Working from home, he applied for and was accepted as a Tier 2 NHS 'Test and Trace' clinical caseworker with NHS Professionals. However, after six weeks, during which time he only received 3 cases to 'trace', two of which were escalated to Tier 1, he took a six-month contract with a local authority dealing with food complaints and undertaking hygiene inspections. Then, when a 'spike' in cases were reported to his authority, he used his training with NHS 'T&T' to good use following up four cases in one day.

Although responses were received from EHPs working in local authorities in west Africa, by some margin more came from South Africa, where local government seems to have been quick to recognise the seriousness and urgency of the matter. As a result, several EHPs indicated that they were working purposefully on COVID-19 control work from February and early March. There is no sense from these respondents that there was anything necessarily preventing them taking the initiative or impeding them from reacting as they saw fit.

The picture that has emerged is of practitioners in countries with communities and populations suffering the poorest standards of housing, sanitation, and water quality, and where there is continuous need for vector and disease control, finding themselves swiftly redeployed and refocussed on COVID-19-related matters. It was as if practitioners whose work normally found them dealing with the most fundamental elements of Public Health were able and capable of being redeployed more readily to the equally fundamental elements of disease control – outbreak investigation, contact tracing, advising affected persons and enforcing quarantine.

3.7 How COVID-19 caused environmental health to re-think its priorities on practice

Without a prepared and rehearsed emergency plan to hand, the appearance of SARS-CoV-2 caused managers to have to make decisions swiftly with little time to draft detailed protocols and comprehensive strategies, instead, as some said or suggested, 'making it up as they went along'. Getting the balance right between continuing to deliver a routine service, and dealing with the infection risk behind the emergency is reported to have been difficult, but there is little sense of panic or an unwillingness to make whatever changes were necessary.

For those in managerial or academic posts, their work initially seems to have been concentrated on developing procedures and protocols for the investigation of outbreaks and surveillance and contact tracing, though some sought to be actively involved thereafter in disease control work themselves. For managers there was the need to consider the safety and welfare of their staff, and for the academics, there was a call for them to develop guidelines for their own institutions on communicable disease control, and to adapt their teaching to reflect the fast-moving developments of the pandemic.

Elsewhere, the response seems to have been more measured and driven by legislation and guidance handed down by central government. Since this was largely focussed on restricting transmission of the virus in commercial premises, and therefore seen as an extension to their statutory health and safety role, this is where EHPs seem to have first been deployed. For those normally employed in food safety, the focus was on how food businesses might make the transition into 'lockdown', and so advising catering outlets, for instance, how they might go from table service to takeaway, home delivery and less commonly, manufacture for online sale. Still other EHPs involved in residential and environmental pollution control found themselves having to consider how to fulfil their purpose when unable to carry out physical inspections of dwellings and locations affected by, or the cause of, pollution episodes. We shall be returning to this in Chapter 4.

3.8 Discovering 'new ways of working'

To a greater or lesser extent an issue for all practitioners and their managers was how they themselves should stay safe. In most cases this was decided by their employers, though one encountered considerable independence of thought on the part of some EHPs. So, where it was decided to close municipal offices and for field staff to 'work from home', it soon became clear to managers that they would be unable to continue to be involved in close management of their staff in terms of the work they performed, and how they went about it.

At this point something remarkable seems to have taken place as EHPs found themselves dominion over their workload, and in what practitioners have remarked was short order, a 'new way of working' evolved. There may have been resistance to this from both sides, but we failed to detect it; rather, the contrary seems to have been the experience, with practitioners finding it liberating, seeing their professionalism recognised.

Free to make decisions as to where best to concentrate their efforts, and now required to look to novel means of establishing the nature of hazards reported, EHPs were able to use their risk assessment skills to deduce the extent to which uncontrolled risks might be set aside, and ones that genuinely required immediate attention. This seems to have been a happy release from meeting targets and proving a more efficient use of their time. However, while some acknowledged that it did create difficulties, many EHPs seem to have found ways to substitute face-to-face interactions with business managers, residential landlords and complainants, with telephone or video contact.

Some agile local authorities clearly responded very swiftly by assessing what was now important about their routine enforcement work, and what might be left to one side, at least for the moment. Then – and this is the most encouraging aspect – EHPs, given a freeing of the reins, took on the challenge and, true to their solution-focussed training and a sense that they could do genuine good, went forth and combined their important regular duties with their new-found public health role. In not all cases, but certainly in some, they seem to have rediscovered their purpose, and with it, a sense of achievement.

Although examples of this now abound, one stood out as deserving of mention, being an early case study on the Local Government Association website in the 'COVID-19: good council practice' section (LGA, 2020), which celebrates some of the innovative ways that local authorities had found to address novel problems.

Assembled in May 2020, 'A district response to COVID-19 (Watford)' sets out the stages of Watford Borough Council's COVID-19 response. The account describes how the managing director, Donna Nolan, one month into her appointment, found herself initiating an urgent review of the Council's business continuity plan. This resulted in the setting up of a Senior Leadership Team that identified the need to 'reset' the Council's roles and responsibilities in the face of the virus and led to a radical new way of operating.

The Team duly recognised, among many other things, the need for '...greater flexibility and adaptability' and recognising '... that new teams would emerge (not just within the council)', allowing '... staff to flourish [since] previous hierarchies might not work, and new organisational "stars" might emerge'. Accordingly. the Council organised itself into 'cells', each with an appointed lead and team, which could focus on the areas of work critical to the council's response and was '... beyond business as usual'.

The case study goes on to recognise the value of operating close to the heart of local communities, regarding the Council's Public Health 'cell' as '... an outstanding example of how a district can make a real difference where it matters the most', in particular, citing the environmental health team who '... opted not to work from home but rather to work frontline in the town to provide the professional advice businesses, particularly shops, needed to open safely'. Aside from 'face-to-face' visits, practical help was provided through COVID-19 information packs and signs designed in-house, the thinking being that people would see the town responding in unison – a 'one town' response.

The report concludes that staff have been '… the key to Watford's COVID-19 journey' by rising to the challenge despite being '… disconnected from their usual working environment and teams [and] facing both professional and personal challenges'. Donna Nolan comments, 'Our strength comes from working alongside and understanding our communities as well as our ability to mobilise quickly and adapt to fast changing situations', suggesting that the town has lived up to its motto: 'To Go Boldly'.

Reflecting on the situation several months later, in an article that appeared in 'The MJ' on 5 August 2020 entitled 'Learning through crisis' (https://www.themj.co.uk/Learning-through-crisis-/2183), Donna Nolan mentions again the creation of '… new cross service working units to deal with the crisis' and the importance of '… good continuity of service via remote working', while recognising the value of the 'Public Protect' team deciding not to work from home in order to '… get out where they were needed most, in the heart of our community, to support businesses …'.

A feature of the various 'new ways of working' described by the respondents was the extent that EHPs found themselves either drafted or finding themselves otherwise joining up into multi-disciplinary teams, often with service providers and individuals with whom they had no previous contact. Here they felt useful both in terms of offering their technical expertise, and for simply having the 'communication skills' to make things work (SM59-Jersey). This seems to have been an experience shared by many EHPs around the world, and one frequently mentioned as an especially positive feature of the pandemic as it panned out.

3.9 Taking a world view on the environmental health response to COVID-19

In addition to personal accounts, we sought impressions of how varied and multi-dimensional the work performed by environmental health service providers had become during the early months of the pandemic from reports, reviews and updates produced by some of the professional bodies and associations identified in Chapter 6. Some of these reports were released mid-year by the national bodies to the International Federation of Environmental Health for the same primary purpose as this book was commissioned: to share and learn from the experiences of others. Material that proved of especial value came out of papers appearing in the academic literature, several being collaborative ventures between academics and EHPs working in the field.

3.9.1 Australia

The response in Australia is especially interesting for the different sorts of activities performed by EHPs in different states and territories. Whereas we feature here what the Environmental Health Australia said in its 'COVID-19 Pandemic Report' (EHA, 2020) we are indebted to Professor Kirstin Ross through personal communication for clarification on matters arising from this Report and for her contribution of insights into the Australian response used elsewhere in this book.

Although normal work in food, water and environmental safety continued into the period of pandemic, state government and local governments were increasingly using EHOs for 'pandemic specific duties', generally in support of State 'action-plans'. The report details, by way of illustration, the activities performed by EHOs [preferred title in Australia] in different states and at different tiers of government working at state level down to local Public Health/Environmental Health Units. It suggests that by June there was much diverse COVID-19 activity underway with significant Environmental Health involvement, describing EHPs '... developing and implementing a COVID-19 response and action plan' [Victoria]; teaming up with 'partner organisations' including the Police [Western Australia]; and co-ordinating 'Incident Management Teams' [Queensland].

In addition, EHOs are described as participating in such things as: border screening and assessment, where, in Queensland, they have been involved in '... greeting arrivals, answering any questions and ensuring they are put onto the right buses' (PC12), assisting State Health departments in case management and contact tracing, issuing quarantine and isolation notices to cases and contacts, and undertaking compliance monitoring and surveillance operations with regulatory partners.

Much is made around the world in praise of Australia's mandatory quarantining of those entering from abroad, and the Report refers to EHOs 'supporting' accommodation providers. Enquiring further about the nature of this 'support', and whether 'hotel quarantining' was still being strictly practised, we were told that it was. Arrivals from overseas go into a hotel or similar facility which is overseen by EHOs.

Here, they are held securely, with guards on each floor, and tested three times – within 24 hours of arrival, on Day 5 and Day 10. If they test positive, they are moved to a 'medi-hotel', though, generally, EHOs are not involved in the management of these facilities. Inter-state migrants from states with high cluster numbers are ordered to self-isolate (PC12).

For Local Government, trying to pursue 'business as usual' as far as possible, and seeking to continue to oversee the control of risk from food, it meant making alterations to inspection regimes, wherever possible conducting these '... by phone with minimal or no site inspections' [Tasmania and Western Australia], whereas, in New South Wales, by June, EHOs were working more in an advisory capacity in support of specialist teams, and seeking to ensure '... uninterrupted environmental/public health and immunisation functions', including '... advising food and other businesses on compliance with COVID-19 restrictions'. Picking up on the involvement of EHOs in immunisation work at his time, we were informed that in most local government areas vaccination programmes came within their ambit, though, until now, focussed, as in Tasmania, on Influenza (PC12).

It was interesting to read of EHPs in South Australia carrying out 'welfare checks' on those registered with the Red Cross as self-isolating, especially of them supporting staff involved in COVID work or work affected by the pandemic, including checking on '... personal impacts of COVID-19 measures on work/life balance ... maintaining regular check-ins to all staff working remotely ...', and '... implementing staff wellness activities, including virtual yoga, meditation and informal team catch-ups via MS Teams or Zoom'.

In a reflective piece by Ingrid Johnston of the Public Health Association of Australia, she observes, 'In many ways, especially in the early phase, Australia's response has been exemplary', containing case numbers when elsewhere they were increasing exponentially, though being an island and able to restrict international travel this was contributary to their success. However, prior to the pandemic Australia had undertaken 'major outbreak exercises ... and response plans formulated and updated as a result', so making it '... as prepared for an outbreak as other countries (such as the UK and the USA), the major difference being that when the outbreak happened, the plans were activated' (Johnston, 2020).

Reflecting on the Australian response, and while not mentioning EHO/Ps by name, Johnston applauds the efforts of: '... community members; health, community and service workers; and national advisory groups and governments', who had '... demonstrated the extraordinary capacity of the community to deal with a significant health threat'.

3.9.2 Malaysia

Choosing mid-year to take stock of the situation in Malaysia, the Malaysian Association of Environmental Health (MAEH) released the 'COVID-19 Pandemic Report' via the IFEH (MAEH, 2020), based on an assessment in early July, when the cumulative data recorded on 7 July 2020 were 8,674 confirmed cases and 121 deaths, and on which day there had been 6 new cases – 2 community-spread and 4 'imported'.

In setting the scene and context for its environmental health response, the Report claims that Malaysia began to make itself prepared in December 2019 on becoming aware of cases of acute respiratory illness in China, aided by their previous experience of Middle East Respiratory Syndrome and the SARS epidemic in 2002/03. At its heart was to have experienced contact tracing teams standing by and preparing Public Health teams ready to screen those seeking to enter the country. Meanwhile, the Government went about upgrading its health facilities and diagnostic capability, roughly doubling critical care bed capacity and the number of ventilators.

When the infection arrived, Malaysia took the radical step of hospitalising all individuals diagnosed COVID-19 positive, whether symptomatic or not, '…' learning from other countries, including China…in identifying do's and don'ts in the COVID-19 response'. By imposing tight controls on those entering the country, just 22 cases were recorded in January all of which were associated with imported cases.

The situation altered suddenly and dramatically on 9 March 2020 when a case in Brunei was attributed to attendance at what the Report describes as '… a mass religious gathering involving more 14,500 Malaysian and 1,500 foreign participants at a mosque in Kuala Lumpur between 27th February and 1st March 2020'. This triggered a series of epidemiological investigations and mass screenings such that by 18 March 2020 the Public Health authorities had established that more than half of the 673 confirmed cases were linked to this event.

With the prospect of an exponential increase in the number of cases, the government decided to impose a 'lockdown', with a phased imposition of restrictions on movement, along with the actions recommended by WHO, from March through to early May. Thereafter '… most economic sectors and activities' were permitted to operate '… while observing the business standard operation procedures', which included social distancing and recording customer details. However, restrictions continued to be imposed in respect of all mass gatherings and sporting activities, and inter-state travel was not permitted except for work purposes.

From mid-June to the end of August, Malaysia entered the 'Recovery' stage, with more sectors of the economy and schools re-opening on condition that they followed relevant SOPs. Events, including those involving large gatherings of people, were permitted, though subject to compliance with SOPs. Key, throughout this time, has been the deployment of environmental health practitioners, of whom approximately 5,000 are employed in the Ministry of Health and 2,000 with local authorities. All are gazetted as authorised officers under the Prevention and Control of Infectious Disease Act 1988.

The pandemic has seen EH personnel involved in a wide range of activities, some of which are detailed in the Report and include screening travellers at borders; investigating reported cases and tracing contacts; monitoring persons under surveillance and self-isolating, supervising the disinfection of public areas, and the burial of those dying from COVID-19. In addition, they have enforced compliance with movement controls during 'lockdown', and taken formal action for offences under the Prevention and Control of Infectious Disease Act 1988.

3.9.3 New Zealand

As early as 14 April 2020 WHO reflected on the value of the early imposition of extreme public health measures, observing,

> Where there has been early action and implementation of comprehensive public health measures – such as rapid case identification, rapid testing and isolation of cases, comprehensive contact tracing and quarantine of contacts – countries and subnational regions have suppressed the spread of COVID-19 below the threshold at which health systems become unable to prevent excess mortality.
>
> World Health Organization (2020)

This, they point out, is not only vital in preventing the direct harm from the virus, but allows countries to maintain high-quality clinical care, reduce secondary mortality due to other causes, and, importantly for professions such as environmental health, allow '... continued safe delivery of essential health services'.

New Zealand's ability to initiate and maintain tight border controls and a robust system of case detection has been held up as a model of what can be achieved by a liberal democracy that might baulk at the Draconian measures seen in eastern Asia. Invited to provide an assessment of the situation mid-year, a report – the 'COVID-19 Pandemic Report' presented by the New Zealand Institute for Environmental Health (NZIEH, 2020) draws attention to the Government's public health strategy, still focussed on 'Elimination' as of 29 June 2020. However, the Report is at pains to point out that 'Elimination' does not mean eradicating the virus permanently from New Zealand; rather, it is when we are '... confident we have eliminated chains of transmission in our community for at least 28 days and can effectively contain any future imported cases from overseas'.

Explaining how the Government expected to have to see the control measures applied indefinitely or until the epidemiology suggests that the disease can be managed through effective treatments and vaccination, it makes clear that their success stems from applying the tightest controls at borders and so preventing new cases being introduced from overseas. The policy, at the outset, of requiring everyone entering New Zealand to be isolated in a Government-controlled facility for at least 14 days on arrival had achieved its purpose.

Thereafter, the NZ system harnessed the two most effective tools of infectious disease control – robust case detection and surveillance – testing anyone with respiratory symptoms and 'sentinel testing' of the wider population to detect asymptomatic individuals, though with particular attention to those who might be, as the report describes '... disproportionately affected by a widespread outbreak', here mentioning Māori and Pacific populations and those in 'institutional settings'.

Recognising the importance of effective contact tracing – minimum of 80% of contacts of COVID-19 test-positive cases traced and quarantined within 4 days of exposure – in preventing onward transmission, the Report makes reference to this being supported by Public Health units following up cases and identifying contacts and clusters duly '... funded to enhance their ability to do so'.

At the time of concluding the drafting of this book, the outcome of failing to follow the cardinal rules of infectious disease control can part be measured by the data collected on new cases identified, hospital admissions and, sadly, deaths. When a 'second wave' of more infectious variants served to slow the decay of their epidemic curves, countries around the world are faced with many of the same choices that they had faced in March 2020, save for notable cases like New Zealand, that have enjoyed sustained respite, thanks in no small part to maintaining strict border controls (New Zealand Immigration, 2021).

A timely article in the CIEH 'EHN Extra' on 29 October 2020 entitled 'How New Zealand dealt with COVID-19' (CIEH, 2020), based on an interview between Katie Coyne from the CIEH and Tanya Morrison, national president of the New Zealand Institute of Environmental Health, talked of the population of five million enjoying a spring in 'near normality' while Europe entered a further phase of restrictions. With just 2,000 cases and 25 deaths, Ms Morrison describes NZ's approach – 'go hard and go early' – a bold decision given all of the unknown, and where their 'lockdown' received the support of the NZ Police to enforce self-isolation and restrict movement outside of the local area or 'bubble', supported by staff of the Ministry for Business, Innovation and Employment monitoring compliance in businesses permitted to carrying on trading.

While communicable disease control work – contact tracing, border surveillance and COVID-19 management – was predominantly undertaken by health protection officers (HPOs) working with the district health boards under the Ministry of Health, EHPs were only called on to assist 'when case numbers were high and extra support was required'. On the occasions that they did, they were involved in investigative work, supervision of disinfection and 'incoming passenger surveillance'. On reflection, 'Effective contact tracing was key in limiting the spread of COVID-19, particularly within the community', and with it, the successful containment of the disease.

In the main, environmental health staff used the time under heightened alert working from home on reports, rescheduling inspections and generally preparing for the moment when they could return to 'face to face interactions', though many were also '... on stand-by and awaiting call out by Civil Defence if required'. Nevertheless, even during 'lockdown' some EHOs responded to requests from district health boards to respond to complaints if it was felt that there was 'immediate risk to public health', for instance, food complaints, surcharging sewers and potentially toxic discharges.

There was welcome openness about there not being '... a nationally consistent approach with regards to the utilisation of all Environmental Health staff' months into the pandemic. How, and to what extent, EHOs were involved in COVID work came down to where they worked, and at, then (29 October 2020), Alert Level 1, not many were '... involved directly with the ongoing management of COVID-19', so now '... the day in the life of an EHO is predominantly the same as pre-COVID times'.

3.9.4 Africa

In a report compiled by the chairperson of the IFEH Africa Group (Chaka, 2020) a surge was considered more than likely than not in some regions despite the best efforts of the public health services in the respective countries. Setting aside the difficulty of speaking for a continent of 44 countries made up of diverse communities, each with their own economic and health challenges, Jerry Chaka reports environmental health professionals being at the forefront of communicable disease prevention, their prominence due, he believes, to their '... active and visible role in COVID-19 projects and programmes'.

The author describes the role of 'outbreak response teams', comprising environmental health professionals working in multi-disciplinary units at district, provincial and national level, where communication is seen as a key element of this system working as effectively and consistently as it might in each country. The plans drawn up by these teams have sought to cover off all elements of the disease control system, from testing for the virus, contact tracing and listing, isolating and quarantining 'persons under investigation', and then on to establishing and monitoring quarantine sites. EHPs in many African countries have been heavily involved in contact tracing and ensuring self-isolation, and on occasion where there has been a real risk this might not be enough to prevent onward transmission, referring the individual for isolation at a government facility, over which some have had managerial responsibility.

Efforts to prevent infection being further imported from abroad has seen EH professionals actively involved, and in some countries leading, the screening of travellers at ports and airports, ensuring that cases remain in isolation ahead of a test for the virus. In addition, they have been posted to health institutions to conduct temperature screening of those attending, again, isolating people ahead of securing a test result, and helping to raise the awareness of public as to the signs and symptoms of the disease.

On the wider issue of health education around COVID-19, African EHPs are said to have played a major role in delivering awareness programmes to communities, in addition to distributing 'Information, Education & Communication materials'. The communities targeted for education are described as 'high-risk communities' such as 'informal settlements', and that during these events attendees would be advised about the availability of other local services, if necessary, arranging referrals for assistance. In addition, the concerns of the communities were recorded, and future programmes adjusted to reflect these concerns.

Other activities mentioned in the report included performing, managing or overseeing the disinfection of outdoor spaces (where often the concern had been the operative's safety), and the enforcement of COVID-19 Regulations in collaboration with the police and sometimes, in what the author describes as, 'blitz operations' led by EHPs.

The Report concludes that COVID-19 has imposed a 'huge burden on Environmental health professionals in Africa', where understaffing continues to impact negatively on the delivery of environmental health services, since '... all human resources are directed towards the prevention of the spread of the pandemic'. It calls on African governments to invest in improving staffing levels to the recommended WHO ratio of 1:10,000 population.

Following on from this, an article for the Western Cape Municipal Brief (2020) reported on a provincial webinar presentation session on 30th September, when Mr Sikhetho Mavundza, Acting Portfolio Head of Municipal Health Services for SALGA [South African Local Government Association], confirmed that rather than 1:10,000, the Western

Cape was operating at roughly half this ratio with 335 EHPs for a population of 7 million, pointing out that they had been providing a '... COVID-19-related service in addition to the usual services that are needed by the communities'.

Outlining the work being undertaken by EHPs at different levels of governance, it was said by one attendee that they were 'first to feel the brunt of COVID-19 in that they formed part of the Response Team and participation in multi-stakeholder engagements', causing them to feel '... burnt out very early in the pandemic'. However, despite this, and experiencing personal risk when working in 'vulnerable areas', they had '... shown courage in the face of adversity'.

In a paper for BMJ Global Health, Morse et al. (2020) consider how EHPs had contributed to the control of COVID-19 in sub-Saharan Africa, suggesting that in many ways '... the continent had never been better prepared to deal with a global pandemic', having experienced major viral epidemics such as Ebola in 2014–2016 before. However, they point out that some COVID-19 mitigation measures such as social distancing and 'lockdown' are simply not feasible either physically or economically, making it imperative to focus on '... context appropriate preventative measures to minimise the impact of SARS-CoV-2 transmission'. These matters are covered in four sources cited by the authors to which we would direct our readers.

Listing many of the key measures identified by Jerry Chaka, the Morse team suggests that as the number of cases rise, there is a case for EHPs assuming a different role and seeing a wider suite of interventions implemented to prevent disease, with these taking into consideration the five interlinked pillars of environmental health created by the African Academy of Environmental Health covering food safety, pollution control, occupational health and safety, built environment and community heath.

They go on to argue that while EHPs have expertise in these five areas, it is '... in their ability to see where these areas intersect, and where an intervention in one area can lead to maximum impact across the whole population' that they come into their own. By way of simple illustration, they offer the situation of a household needing water and having to consider drawing this from a communal point '...where it might be difficult or impossible to practice social distancing or disinfection of hand contact surfaces after each use ...'. Finding a way of managing and so mitigating this risk '... requires a holistic view which can coordinate a transdisciplinary response engaging both experts and community members'.

They conclude,

> Issues surrounding the control of infections such as SARS-CoV-2 are complex and wicked in nature. To tackle current and future

pandemic situations, we require to take a transformative approach to preventive health, and in a world where the majority of qualifications are now narrow and specialised, EHPs bring a broad spectrum of skills which allows them to 'dance across disciplines.'

This idea of EHPs being able make frequent and significant changes of course to meet need and demand came out strongly in our research. Taking Africa as a whole, the initial efforts of EHPs seem to have been concentrated on local contact tracing and delivering advice and guidance to the public in various settings, for example, schools and community meetings. They, then, seem to have moved on to ensuring compliance in commercial premises with COVID requirements ahead of closure under order of whatever national 'lockdown' mechanism was implemented. Thereafter, it was about monitoring restrictions on gatherings, like weddings and funerals, and undertaking decontamination and sanitation procedures, with some, then, being assigned to overseeing quarantine arrangements and the storage and disposal of the dead.

An article that appeared in the *North Glen News* (2020) ahead of World Environmental Health Day on 26 September 2020 in which the health minister, Dr Zweli Mkhize, described, in a keynote address via webinar, how environmental health practitioners have played a critical role in the outbreak of the pandemic in the country. He draws particular attention to their work at points of entry into the country where they had '… the unenviable task of preventing the possible importation of Covid-19 as the first line of defence', not only to protect South African citizens but many other African countries to which the country was a 'transit hub'. He concludes that these measures '… helped to not only delay the introduction of the disease but also an opportunity to strengthen its preparedness of the inevitable entry of the disease into [the] our country'.

References

Chaka, J. (2020). *Role of environmental health practitioners in the prevention and control of COVID-19 disease: The African perspective*, posted on IFEH website (https://ifeh.org/) on 11 July 2020. Available at: https://ifeh.org/covid19/docs/CORONA_VIRUS_IN_AFRICA_IFEH_Africa_Report.pdf (accessed: 11 February 2021)

CIEH (2020). *How New Zealand dealt with COVID-19*, EHN Extra (online), CIEH. Available at: https://www.cieh.org/ehn/public-health-and-protection/2020/october/how-new-zealand-dealt-with-covid-19/?utm_campaign=11910921_EHN%20Extra%2029.10.2020&utm_medium=email&utm_source=CIEH&dm_i=1RSV,73AIX,OPDOS2,SNMM3,1 (accessed: 22 February 2021)

Environmental Heath Australia (2020). *Environmental Health Professionals around Australia – COVID-19 Pandemic Report*, made available via IFEH (https://

ifeh.org/) at: https://ifeh.org/covid19/docs/Environmental%20Health%20 Professionals%20Australia%20-%20COVID-19.pdf (accessed: 11 February 2020)

Johnston, I. (2020). 'Australia's public health response to COVID-19: What have we done, and where to from here?', Commentary piece for the *Australia and New Zealand Journal of Public Health*, 44(6), 440–445. Available at: https://online library.wiley.com/doi/epdf/10.1111/1753-6405.13051 (accessed: 30 December 2020)

Local Government Association (2020). 'COVID-19: Good council practice' section on LGA website. Available at: https://www.local.gov.uk/covid-19-good-council-practice (accessed: 28 October 2020)

Malaysian Association of Environmental Health (2020). *COVID-19 Pandemic Report* – Prepared on behalf of the Environmental Health Officers and Assistant EHOs of Malaysia and made available by the IFEH at: https://ifeh. org/covid19/docs/Environmental%20Health%20Officers%20and%20Assistant%20Environmental%20Officers%20Malaysia%20-%20COVID-19.pdf (accessed: 3 December 2020)

Morse, T., Chidziwisano, K., Musoke, D., Beattie, T.K., & Mudaly, S. (2020). Environmental health practitioners: A key cadre in the control of COVID-19 in sub-Saharan Africa. *BMJ Global Health*, 5, e00314. https://doi.org/10.1136/bmjgh-2020–003314. Available at: https://gh.bmj.com/content/bmjgh/5/7/e003314.full.pdf (accessed: 11 February 2021)

New Zealand Immigration (2021). *COVID-19: Key updates*. Available at: https://www.immigration.govt.nz/about-us/covid-19/coronavirus-update-inz-response#travel-to-new-zealand (accessed: 10 February 2021)

New Zealand Institute for Environmental Health (2020). *COVID-19 pandemic report* – Presented on behalf of Environmental Health Officers and Health Protection Officers New Zealand – and made available through the IFEH website. Available at: https://ifeh.org/covid19/docs/Environmental%20 Health%20Officers%20and%20Health%20Protection%20Officers%20 New%20Zealand%20Covid%2019.pdf (accessed: 14 February 2021)

North Glen News (2020). Keynote address via webinar of health minister on the occasion of World Environmental Health Day (26 September 2020).

Western Cape Municipal Brief (2020). Article: *SALGA Western Cape hosts the provincial webinar on Municipal Health Services* (28 October 2020).

World Health Organization (2020). *Covid-19 Strategy Update – 14 April 2020*. Geneva: World Health Organization. Available at: https://www.who.int/docs/default-source/coronaviruse/covid-strategy-update-14april2020.pdf?sfvrsn=29da3ba0_19 (accessed: 10 February 2021)

4 How environmental health practitioners responded to COVID-19

4.1 Introduction

The work of environmental health practitioners covers many factors that have an impact on health, where we live, the places we work and recreate, the food we eat and, as we have seen through this pandemic, our vulnerability to infectious disease. Here we demonstrate how EHPs employed in both local government and the private sector have set about putting their knowledge, skills, and experience to good use in responding to the issues arising from the infection, especially those that have had a bearing on their routine work.

Based on the contributions received from practitioners working as local and regional government officers, and those acting as consultants and advisers to industry on compliance and health protection, we begin to see a profession that understands its core skills, and, more specifically, the transferability of these skills. EHPs report performing disparate activities, which, under normal circumstances, might not have fallen to them, though now, through necessity and a willingness to 'make it happen', they have made them work, and work well.

Inevitably, then, the focus of their contribution during the first 'wave' was on their role in taking the issue of infection control to the many and various locations where the public might infect each other, or in workplaces, where employees might also be at risk. However, this changed under pressure to keep businesses open for as long as it was deemed 'safe' (and, similarly, allow them to re-open under the same circumstances), thus increasingly focussing on directing and advising proprietors of business premises on restrictions imposed through legislation and guidance.

As has been mentioned previously, the capacity of EHPs to respond to so many aspects of the COVID-19 crisis is widely held to be a product of their training and the experience of discharging their regulatory role sensibly and pragmatically. While they already had a close working relationship with the business communities in their districts, when called

DOI: 10.1201/9781003157229-4

on to advise and enforce COVID-19–related guidance and legislation, this came to the fore.

Here, the activities with which they found themselves engaged are brought together for convenience under the titles used in service delivery. The challenges and obstacles that practitioners encountered, the opportunities that these encounters presented, and how they believed their efforts and abilities had served to benefit health and raised the profile of environmental health and the profession, will be considered in Chapter 5.

4.2 Providing guidance and regulatory direction to businesses

The announcement that the spread of COVID-19 across continents was such that it had been up-graded on 11 March 2020 to 'pandemic' meant that it was increasingly likely that restrictions would be put in place on the public entering enclosed spaces to work, making purchases of non-essential goods, providing 'close contact' treatments and enjoying hospitality. Inevitably, then, as soon as the governments of the world began passing legislation and drawing up guidance setting out these restrictions, it would fall to whoever was considered able and available to deliver this to business operators.

Although different countries have different ways in which they discharge the various functions of environmental health legislation, a core set of services are commonly performed by their central, regional, and local government administrations, namely, food safety, occupational health and safety, housing, environmental pollution control and public health. To this should be added infection and vector control where zoonotic infections are endemic.

Whether working as a specialist or as a generalist in one or more of these fields, the appearance and rapid increase in prevalence of COVID-19 found practitioners having to give an increasing proportion of their time to these more pressing issues at the expense of their routine work. Nevertheless, where and how they were re-assigned seems to have taken account of their previous specialism, thus allowing a degree of 'business as usual' to prevail for a short while, even if, by the height of the first wave, a far greater proportion of their time was being dedicated to COVID work.

That said, the period most spoken of as especially demanding in terms of interacting with businesses came, not surprisingly, as countries came out of 'lockdown', with restrictions being progressively relaxed, and shops, schools, markets, and sports and performance facilities re-opening. EHPs were now in the sometimes-difficult position of 'supporting safe transition' and avoiding this being responsible for future outbreaks. In Australia, for instance, inspections were undertaken to ensure that a 'COVID-19 safe plan' was in place and being implemented against their wider 'business continuity plan'. Where this was not the case, EHPs would work with the business to produce one, offering training and education to their staff.

Throughout, it was evident that EHPs were employing their capacity to win over reluctant business proprietors by their powers of persuasion to good effect. However, on occasions, they clearly had to resort to formal action through the service of prohibition notices, and in one case emerging from our study, a local authority having to prosecute a tanning salon that failed to close in the first lockdown in England after repeated warnings and the service of a notice.

4.3 Delivering information and advice to the public

Perhaps one of the clearest differences between the responses to COVID-19 in countries around the world has been the degree to which EHPs have been actively involved in preparing and delivering public service information. One factor might be the extent to which national governments wish to maintain custody over the message so that they can exercise control over its communication in a timely fashion. It remains to be seen how effective public information of this sort can be, especially if it must be sustained beyond 12 months.

Where they have been allowed to do so, EHPs have shown that they are a good local resource for getting information to the most vulnerable, who may not be known to other health agencies. One is struck by the way in which EHPs in South Africa were deployed to take the hygiene message out beyond the commercial sector and into communities. Here, they felt free to use their transferable skills to good effect in a wide range of places where people might be expected to gather, or otherwise especially exposed to infection.

4.4 COVID-19 and food safety

EHPs that would normally be undertaking routine food safety work – essentially one or more of the core elements of inspecting premises involved in primary production and food manufacture; advising food business operators on any food risk assessment tool used to determine where control measures should be applied such as 'Hazard Analysis Critical Control Point (HACCP); and enforcing hygiene legislation designed to protect the final consumer – felt able to set more or less of this aside.

It is not immediately apparent from our survey how much of their reactive work, for instance, the investigation of food complaints, urgent hygiene issues such as pest infestations, and follow-up on surveillance evidence of a food-borne infection, was 'put on hold'. However, several respondents said that these services were maintained, but only after strenuous 'risk assessments' had been performed. It also remains unclear what the consequences of this necessary prioritisation have been in wider public health terms.

The policy of instructing people to stay at home and keep interactions with others to an absolute minimum meant that hospitality premises had to close, while allowing, as we have said before, food business operators to prepare and serve food through takeaway or home delivery services. Consequently, EHPs – both those working in the public sector and commercially – found themselves in demand for advice, guidance and, when required, re-registration. In addition to supporting established food businesses to diversify, there was unexpected demand on EHP time from those with no previous experience in commercial food preparation being put on furlough or facing redundancy deciding to set up food businesses with little understanding of the hygiene implications,

Under normal circumstances this might have meant careful consideration of new and revised HACCP plans and alteration of registration details, but time was in short supply in March and April. Thus, EHPs reported that they had found ways of being able to perform this without the need for inspection and face-to-face meetings, though offering close support to those new to the industry. In Australia, for instance, quarantine instructions meant that all food businesses had to shift to a takeaway service, though EHPs continued in some states to undertake inspections 'virtually'.

A further unexpected call, though none more urgent or important, was support for charities and ad hoc groups of individuals looking to supply food to the poor, 'shielded' or otherwise vulnerable during 'lockdown'. Those EHPs finding themselves advising the volunteers behind these initiatives were faced with a difficult task; consider the good they were doing to relieve the plight of those unable to shop and feed themselves or their families against the risk of contracting a food-borne disease, while all the while considering the overall risk to all from the virus.

4.5 COVID-19 and occupational health, safety and hygiene

Although it is by no means the case universally, local government and thus many EHPs are responsible for enforcing health and safety legislation in the workplace. In the UK, a parallel system of safety enforcement exists between a central government agency – the Health and Safety Executive – and local government, both working through the Health and Safety at Work etc. Act 1974. This Act extends protection to those who might be affected by a business operation other than their employees, thus the public at large.

While 'lockdown' would provide the near total elimination of the risks posed by a business when fully operational, COVID-19 raised fresh challenges for those seeking to protect their staff and customers. Thus, for instance, in New South Wales, Australia, EHOs with expertise in water and sanitation matters found themselves extending the COVID-19 message to utility providers, thereby supporting their local government colleagues.

For the most part EHPs normally involved in occupational hygiene and workplace health and safety found that their 'routine' work evolved to meet the problems thrown up by COVID-19, so SM62's health and safety enforcement role in the UK was largely exchanged for 'COVID-compliance' work centred on an office dedicated to enquiries, complaints and enforcement of relevant legislation, an arrangement that seems to have been followed by local authorities elsewhere in the UK and beyond, including the United States (SM60).

Interestingly, COVID-related occupational health and safety was encountered first by EHPs when 'panic buying' denuded the shelves of supermarkets and convenience stores of the usual commodities when disaster looms: flour, canned vegetables and toilet-rolls! This would have been no more than a nuisance if the event that triggered it had not been associated with a respiratory infection transmissible through droplet and aerosol dispersal that immediately called for 'social distancing' and the need to disinfect hands and hand-contact surfaces. Although the major food outlets were quickly onto this, the smaller ones placed a considerable strain on practitioners employed in both the public and private sectors. It required of those overseeing the management of the risk the imagination to find new ways of interacting with businesses, and the adoption of a pragmatic approach that recognised the need for greater emphasis on encouragement rather than enforcement.

Clearly, where exemptions were granted to re-open on condition that the businesses met certain minimum standards of personal protection, physical separation, and hygiene, EHPs were the natural choice to interpret the guidance and provide practical advice, given their deep understanding of the means by which the coronavirus was transmitted. In this case officers that were authorised to enforce health and safety legislation could make their visits 'dual-purpose', though one practitioner

(SM62-UK) observed that there was less routine work to do anyway which they put down to there being far less hazardous activity going on because of 'lockdown' and suggesting that businesses were being 'more careful'.

A rather different experience was reported by an EHP (PC7), also in the UK, who described the gradual discontinuation of their work prior to and during 'lockdown', the challenge really coming with the imminent emergence from 'lockdown' in his inner-city authority:

> Attention turned to the risk of legionellosis from showers in reo-pening gyms and leisure centres, but especially when 'close contact services' (hairdressers, barber shops and beauty salons) recommenced business. At first, he recalls, there was misunderstanding over the requirements regarding PPE, but then the staff seemed to grow tired of donning it. Whilst it was not such a problem when the premises were small in size – in which case he was able to ascertain compliance 'from the doorway' – on several occasions he had cause to ask people to stand away from him during visits.
>
> As the weeks went by, he noticed businesses becoming more and more confused by the legislation and guidance, and while never openly aggressive or threatening towards him, they became more argumentative, venting their frustration at Government policy, the council being seen as one and the same. Several businesses raised the behaviour of Dominic Cummings [the Government's chief policy adviser who travelled to Durham having tested positive for COVID-19 in alleged contravention of the law] as an argument against criticism of their own conduct.
>
> Seeing how businesses responded to the heavy-handed approach adopted by others in pursuing COVID-19-related matters, he and his colleagues chose to follow a 'softly-softly' approach, explaining things calmly and in simple terms. This served to defuse difficult or awkward situations, and while their preferred approach under normal circumstances, it was especially effective when it came to COVID-19. This preparedness to adopt 'the common touch' was what he felt marked EHPs out.

In the knowledge that food retail outlets would be allowed to remain open to trade throughout any period of 'lockdown', EHPs, both in local government and working in an advisory capacity in the private sector, found their awareness of the principles of cleaning and disinfection in demand. However, the objective of '… creating an environment where everybody was COVID-19 prevention compliant' (SM30-South Africa) at short notice and at pace was hampered by simply not having enough staff to call on.

Finally, a mention was made previously to EHPs in Australia supporting colleagues in local government to maintain their physical and mental well-being, but programmes were also developed for the wider community, including initiatives around health and wellness that included 'virtual physical, yoga and meditation training'. (PC12)

4.6 COVID-19 and housing

Without doubt, the pandemic has highlighted the contribution made by EHPs in fulfilling the objectives of Public Health, while seeking to maintain economic activity, something that has become known during the pandemic as 'protecting lives and livelihoods'. Rather less obvious, though no less important, has been their contribution in dealing with housing matters aggravated by the pandemic.

Here we encounter one of those points of diversity between the countries of the world, like the UK, where EHPs perform a central role in housing, and the many others where EHPs only deal with housing tangentially, such as through drainage or pest management issues. Yet the pandemic, rather than masking the association between housing conditions and public health, has thrown it into sharp relief, and, with so much time spent at home, confirmed the domestic environment as a determinant of health.

What has become apparent during the first nine months of the pandemic is how it has affected populations disproportionately and, if the plight of those living in poor quality housing had not been appreciated before, it now shone a light on health inequality and inequity. Even before COVID-19, Michael Marmot and his Institute of Health Equity team at UCL, revisiting the Marmot Review of 2010 (Marmot et al., 2020a), found that housing conditions had deteriorated in the UK for many, and in the knowledge that deprived areas were associated with shorter life expectancy, the social gradient had become steeper in the last decade.

In 'Build Back Fairer' (Marmot et al., 2020b), aside from showing how COVID-19 mortality rates follow a similar inequality gradient to non-COVID mortality, it demonstrates that 'overcrowded living conditions and poor-quality housing are associated with higher risks of mortality from COVID-19 and these are more likely to be in deprived areas and inhabited by people with lower incomes'. This, and the disproportionate harm done to BAME people through occupational exposure to the virus in health and social care, transport, manufacture, and so on, only served to emphasise why future public health policy should be framed in the widest environmental health context.

4.6.1 'Overcrowding' and the residential spread of COVID-19

Given that the central plank of most 'lockdown' measures to contain the spread of COVID-19 was, and still is, 'Stay at Home', being confined to cramped, noisy, and poorly insulated accommodation would be potentially damaging enough to physical health and mental well-being, without the added risk of infection caused by multiple occupation, overcrowding, and multi-generational occupancy.

Where the local authority holds the housing remit, EHPs are likely to be aware of the properties that are occupied by more than one household – houses in multiple occupation (HMOs) as they are known in Britain – especially if they are subject to licensing under a registration scheme. Other countries have different descriptions for similar types of properties and shared accommodation; for example, in Australia, they fall within the category of 'rooming' or 'boarding' houses.

The issue of crowding – a feature of older and smaller properties occupied by low-income households, sometimes consequent on high housing costs – has been highlighted during the pandemic. This includes migrant workers living in dormitories in Singapore and people living in multi-occupied accommodation in the UK during the lockdown, where contributory factors in COVID-19 mortality rate in areas with higher levels of deprivation are household crowding and lack of space. This has been shown to be the case with multi-generational households in the UK, where the risk of transmission is greater (Marmot et al., 2020b) and may be responsible for increased severity of illness and poorer resolution, as living in close proximity may increase the viral load transmitted.

While the increased risk of household transmission of SARS-CoV-2 is understood (Denford et al., 2020), it is evident that the risk of 'within-household infection' is greatly increased among those from low-income BAME communities. This has prompted local authorities to seek to protect these communities by issuing advice and guidance to those in poor housing, especially those renting privately.

4.6.2 Dealing with poor house conditions aggravated by COVID-19

When COVID-19 arrived, housing problems did not conveniently disappear. As a recent report for the Northern Housing Consortium (Brown et al., 2020) observed, many households were living with long-standing repair and quality issues before the pandemic, so 'lockdown' had ultimately '... worsened such conditions and impaired people's ability to live with those conditions'. In addition, there was the problem of carrying out repairs where restrictions would prohibit tradespeople entering the property. There remains the long-standing issue of renters being unwilling to report repairs or press their landlords to remedy defects for fear of eviction, even where the tenants were protected against this by an eviction ban. The report goes on to make clear: '... the COVID-19 'lockdown' has shown in the starkest of terms that run-down homes are resulting in run-down people'.

To assist EHPs in the UK in pursuing interventions on behalf of tenants in this situation, the Ministry of Housing Communities and Local Government issued guidance on enforcement of standards in rented properties (MHCLG, 2021). So, where it would not be possible to inspect a property, for instance, where someone was self-isolating, it might be appropriate to reduce the priority of lower-risk hazards or an assessment could be made through photographs, video or live broadcasting by the tenant, or follow, as one senior manager in the UK put it, 'the art of inspecting, without inspecting' (PC13). Reckoning the risk of not acting on the complaint, against the risk of COVID-19 infection, became a regular consideration, causing local authorities to move to a more reactive model of enforcement based on complaint, often at the expense of pro-active multi-agency operations.

One EHP reported carrying out virtual inspections via WhatsApp video with occupiers showing him around, though there was a concern that gathering evidence in this way might not meet the necessary standard of proof if a decision was challenged. On disrepair complaints the same EHP devised a guide for administrative staff to ask questions about whether anyone in the property had COVID-19 symptoms or was in quarantine because of underlying health conditions, thus putting them at greater risk. By this means it would be possible to gauge whether a potential Category 1 hazard existed (where there is a duty to act), or if it only amounted to a Category 2 hazard, in which case the complainant would be informed that they could not be helped at this time.

One private sector surveyor and chartered EHP, who described most of their work since the first lockdown was lifted as drafting expert witness reports for social housing tenants taking private action against their landlords, said that although they were still carrying out inspections, these needed to be conducted with due regard to social distancing and personal hygiene, and they had moved away from a clipboard and pen, and were now using modern IT equipment (an iPad and Apple Pencil) to take notes. This was especially important since the pandemic had alerted them to '… a lot of really poor housing conditions in the social housing sector' and resulted in them advising tenants to consider '… other options if their private sector housing department is unwilling to act…' (SM61).

4.6.3 Reaching those in poor living conditions challenged by COVID-19

If the pandemic gave local authorities discretion over housing enforcement, it was not held to extend to the most pressing issue of the moment – COVID-19 transmission in shared housing, both multi-generational and those in multiple occupation. Accordingly, we received examples of local authorities that had convened multi-disciplinary teams, in one case bringing together EHPs from the Director of Public Health's staff, the 'Housing Options' team and the Environmental Health HMO licensing team to develop advice and guidance in several languages appropriate to the communities, thus helping those living in houses in multiple occupation to manage the risks from Coronavirus in accordance with the Government's 'Stay at Home' message. This gave advice on what additional hygiene and social distancing measures were necessary where the property was crowded.

Once produced, the local knowledge for which EHPs have been much praised recently in respect of 'test and trace', and especially their knowledge of local housing conditions and the private-rented sector, meant that the guidance could be distributed to all HMOs on the department's register via locally established channels, including landlord and managing agent forums, thereby ensuring that those living in vulnerable circumstances, and in greatest need of advice and support, received it.

In another borough, one of the first things that the manager of the Residential Section did was to write to all HMO landlords (the authority benefited from having a licencing scheme) providing them with advice on how they could help tenants who needed to isolate or quarantine, while further signposting them to government and PHE weblinks. The letter made clear that the authority '… would not tolerate illegal evictions' and that they would expect landlords who were no longer receiving rent because their tenants had had their work interrupted by COVID-19 to negotiate with them alternative payment plans, or else '… accept a lower rent during the period'. These discussions should be held as soon as possible, thus allowing landlords to apply for 'mortgage holidays' (PC13).

Beyond securing tenure, local authority EHPs have used their communications with landlords to urge them to encourage their tenants to advise them of occupants displaying symptoms and to self-isolate for what was then the recommended period of 14 days. Furthermore, landlords and managing agents should seek to further safeguard their tenants by, for instance, reviewing the adequacy of waste collection; enhancing the cleaning and disinfection of common parts; increasing the ventilation of shared facilities; and locating hand sanitisation and posting Public Health England advice around their properties.

4.6.4 Finding novel ways to respond to the impact of COVID-19 on housing

Various ideas have been mooted by the environmental health community over the last months to help support those living in poor housing conditions who find themselves especially at risk from infection, particularly if they have tested positive for COVID-19 and are required to self-isolate or are quarantined as being extremely vulnerable.

Various organisations that are either run by EHPs or have EHPs as key members of their team, including RHE Global (part of RH Environmental Ltd) in collaboration with DASH Services – a joint-working initiative with local authorities, property owners, landlords and tenants – have provided help and advice to local authority EHPs, landlords and tenants during the pandemic. This has included providing advice and template letters on electrical and gas safety checks, and notification of essential disrepairs including when the tenant is self-isolating. RHE have also provided a means whereby EHPs might consult colleagues on problems they have experienced.

An example of the sort of dilemma thrown up by SARS-CoV-2 was highlighted to us by an officer who had been contacted by the landlord of two tenants who had rented separate rooms in an HMO, who were returning from a country from which quarantine was required. Ordinarily this would not have been a problem, except that they were also bringing family members with them, and to stay in the HMO would have breached licence conditions when the advice from the Government was that people that you might be staying with did not need to stay at home '… unless they travelled with you', which, of course, they did. There was also concern expressed as to how this might affect the other residents of the HMO sharing facilities.

Other ideas have had the potential to succeed, being consistent with Government recommendations or that they made rational sense. One local authority had the idea of temporarily re-housing older occupants living in multi-generational households, though this was considered unworkable in practice, not least because of the lack of accommodation and potential adverse impact that this might have on mental health. Pursuing the line of finding alternative temporary accommodation for individuals who might be infectious so as to protect others, the idea of relocating those required to self-isolate – a coherent strategy applied in Australia to Australian citizens arriving home and required to quarantine for 14 days being put up in hotels – seemed only to come down to a lack of finance.

Finally, we should mention a private sector chartered EHP (SM32-UK) who, from the very start of the pandemic, and on an ad hoc basis, was '… advising the lettings industry on regulatory compliance'. Thereafter, they had been '… cascading [down] information via a website, and encouraging local authorities to take a pragmatic approach to regulatory compliance' as the letting industry shut down in the first 'lockdown' and property inspections were paused. This had resulted in them '… standing back and seeing the overview, analysing issues from different perspectives, seeking solutions'.

4.7 COVID-19 and environmental protection

As with other environmental health issues, the arrival of the infection in countries around the world would not be cause for nuisance or environmental pollution hazards to diminish in number or severity; indeed, as we shall see, they became more apparent, especially during 'lockdown'. However, initially, interventions seemed likely to be set aside in favour of the more pressing need to respond to the immediate risk of infection, though for some EHPs normally involved full-time in pollution control work, this meant allocating time in their day for 'COVID-work'. One English EHP (SM26), normally involved in pollution control–related work, talked of being assigned other duties involving the delivery of medicines and food, followed by some 'test and trace' work, but emerged feeling that they had been 'undermined' by their senior managers.

On the other hand, in the Channel Islands, an EHP respondent seems to have been quickly redeployed from 'pollution regulation' to '… case interviews and contact tracing' and with the success of the local 'test and trace' system '… putting them in a very good place' and businesses re-opening '… routine work has recommenced'.

This apparent preparedness to see environmental protection/pollution and nuisance work – here also thinking of pest and waste management and drainage and water supply issues – set aside in these months leading up to and into 'lockdown', was a source of concern to some EHPs. A public sector EHP in South Africa specialising in air quality and waste management in the District Municipal sector found that as the COVID-19 crisis worsened, routine work was affected adversely and objectives not met.

In the Republic of Ireland, the imposition of restrictions on movement during 'lockdown' forced people to stay at home for what turned out to be three months of unseasonably hot weather from April to June. As described by PC5, EHPs responsible for environmental protection very quickly discovered the consequences of 'dormitory suburbs' becoming workplaces, home schools and play areas, as they experienced an increase in domestic noise and air pollution problems from what was described as 'the informal entertainment sector' due to the enforced closure of hospitality venues, and the burning of uncollected household waste.

The respondent goes on to explain just how difficult this was from the point of view of the EHP, now required to undertake '... an extensive suite of health-risk assessments' before they met anyone, and to obtain 'essential travel' cover letters to pass checkpoints. Added to this, with work colleagues experiencing COVID-related health problems or required to self-isolate, there was a need to re-assign work. The whole thing had served as a 'sharp reminder' that aside from the risk of occupational exposure they experienced '... the same fears and stresses' as members of the public.

Other EHPs reported an escalation of noise complaints during the first lockdown in England, most of these being complaints of disturbance associated with residential accommodation. Such was the extent of the problem – again, attributed to more people staying at/working from home – that officers had to be moved from other areas of work to help the noise team that was unable to cope with the numbers of complaints.

Even before COVID-19, mental health conditions were a source of major concern, being recognised under normal circumstances as the cause of adverse psychophysiological and social effects (WHO, 2018). With clear evidence that a significant proportion of the adult population was concerned or very worried about the effect that COVID-19 was having on their lives, the combination of financial insecurity, poor housing and the irritation of noise could only be expected to exacerbate an underlying mental health problem, especially during periods of 'lockdown'.

In addition, EHPs reported an increase in pest problems believed to be associated with difficulties maintaining routine household waste collection during 'lockdown' and the closure of recycling and household waste centres. When such problems – waste and uncontrolled pest activity – come together in what are known in Britain as 'filthy or verminous premises', they are hard enough to remedy even under normal circumstances. An EHP in an English local authority reported that during the lockdown they had such a complaint and this proved especially problematic given the need to enter the property.

Finally, local authority EHPs in the East Midlands of England were alerted to problems of rubbish accumulating in the rear yard of a two-storey terraced house. The subsequent inspection found seven people living there in five households, with those on the first floor, having no access to the rear yard, forced to throw sacks of rubbish out of the windows. The local authority prosecuted the landlord for offences under the Housing Act 2004, demonstrating how one environmental health problem is often symptomatic of another.

4.8   Involvement of EHPs in local and regional disease control initiatives

Given the fact that the aetiology of microbial disease, outbreak investigation and infection control might be expected to feature in courses and training programmes leading to qualification as an EHP world-wide, deployment or redeployment, to a local or regional Public Health 'hub', would, indeed, should, have been obvious and immediate. In many countries this seems to have been the case; thus, in the United States, we see what appears to have been a seamless move in March from working in food safety and vector control to carrying out '... case investigations and contact tracing ...' (SM36).

Similarly, in South Africa, case investigation replaced general duties and responsibilities with the arrival of the infection, which for one respondent (SM29) meant February, but for others, meant March, when they became involved in contact tracing and education on preventative measures – mask wearing, hand hygiene and social distancing (SM25 and SM35). However, in several cases, this preceded deployment to the disposal of the dead. In the case of Nigerian SM24, a practitioner previously involved in disease surveillance, this led to them leading a team of health workers in the field, before being '... redeployed to commence decontamination exercises' while continuing to be involved in case investigation through a federal medical facility.

4.9   Involvement of EHPs in centralised and local 'test-trace-isolate' systems

A feature of the response to COVID-19 in different countries around the world that should invite future research interest is the extent to which some have sought to establish a centralised system of following up index cases testing positive for the virus, whereas others have relied on existing local systems.

In countries that have been previously challenged by epidemics of viral respiratory infection such that they can 'test and trace' at pace and scale, and crucially, have mechanisms in place to enforce isolation – here, thinking about China, South Korea, and Taiwan – the impact, at least during the first wave, was impressive. However, in the case of countries that missed the opportunity to contain the spread/suppress infection this way, our survey would suggest that most relied on regional and local teams to organise and perform the investigations and contact tracing.

While we would surmise that there were relatively few EH-qualified practitioners working in infectious disease control at a national or regional level in the UK at the start of the pandemic, this changed as the situation developed, and the demands falling on the PHE and their health protection teams to perform containment measures came under strain. However, we heard from one former EHP working as 'Health Protection Practitioner' for PHE (SM23) who soon found themself leading a substantial team of 'Tier 1' practitioners aside from '... dealing with cases and minimising outbreaks', their other work had been '... put on the back burner'.

Another UK EHP (PC2) provided an insightful picture of his experience following his freelance work 'drying up' and his effort to contribute to contact tracing at the time he anticipated EHPs would be called on to help, since '... it was important to me to contribute to this public health crisis'. This is described below from his extended account which began when he signed up to the register of volunteer EHPs being compiled by Gary MacFarlane at the CIEH [see Chapter 6] feeling, as he recalls, 'frustrated for a while by the lack of a "call to arms" for EHOs ... wondering who was organising and/or doing this contact tracing work – was it even happening? What was the hold up?'

While he was in line to be recruited as a Tier 2 clinical caseworker with NHS Professionals, for reasons unknown to him, he was offered a regional Tier 1 post with Public Health England, thus '... helping with more complex C19 investigations', and despite, frustratingly, doing 'numerous short, online induction courses', some of them twice, and finding the training '... too generalised and not that relevant to the role', he was, as he put it, 'let loose'. Here are his observations following the completion of his training, where, he was surprised to find, '... I was the only EHO, the rest being '... predominantly ex-NHS/health workers'.

'As time has gone on and I have witnessed more of PHE's approach, I really feel that the lack of EHOs in their Contact Tracing effort has been a real detriment to their work. Perhaps we take it for granted as a profession, but that ability to communicate with all ranges of people and…businesses, is something I think the more medically minded health professionals are just not as good at.

For example, as a Tier 1 Contact Tracer, there was a great deal of liaison with workplaces and "cases". PHE use the term risk assessment to require their Contact Tracers to check the C19 controls at workplaces. This is something much more familiar to EHOs than it is to the other allied professionals, as is liaison with the workplace enforcement authorities – HSE & LA EHOs.

I think that this lack of understanding amongst the non-EHO Contact Tracers did not help matters, especially in relation to investigations involving pubs, factories, and large workplaces. I also felt that there was a degree of professional snobbery with some at PHE who regard themselves to be a little bit superior to EHOs (echoed by some of the full-time EHOs working for PHE).

I'd like to think that I'm one of the flag wavers for our profession and are showing my PHE colleagues that EHOs have a lot to offer in this C19 crisis. I would like to think that PHE would seek to attract more EHOs to their ranks in an effort to balance the type of professionals they have working for them.

Because of my background I have responded to PHE requests to look at certain/specific areas in more detail. I feel that has helped provide a more balanced PHE approach to matters…'

In England, though not the UK as a whole, once the containment stage had given way to mitigation/delay, and the numbers of cases began rising exponentially, a decision was taken to entrust responsibility for contacting, tracing, and advising on isolation to a central hub. Thus, a major contract was awarded to two companies to build a telephone-based 'test and trace' system, recruiting contact tracers, as mentioned previously, through NHS Professionals. This went 'live' on 28 May 2020, with the promise of a telephone app to support it.

It is not for us to say anything more about the effectiveness of 'NHS Test and Trace' that has not been said in the six months of its existence. What we do question is why 'upper tier' local authorities, with a significant public health resource available under the expert direction of Directors of Public Health, were not mobilised in the early stages to investigate cases and local outbreaks, followed by the staged transfer of potentially thousands of qualified EHPs working in lower-tier district councils joining them as the number of cases rose.

Part of the reason clearly lies in the way in which the measures to control the spread of infectious disease has evolved down the years, and especially the mechanism in place in February 2020 when SARS-CoV-2 is likely to have arrived in Britain. The back-story is complicated, but for reasons that may emerge in time, Public Health England, the executive agency set up to assume the infectious disease control role of its predecessor, the Health Protection Agency, seemed to diverge from the strategic plan set out in the 'Infectious Diseases Strategy 2020–2025', sub-headed 'Addressing urgent threats in the 21st century' published in September 2019 (PHE, 2019).

This Strategy presents a comprehensive blueprint for managing future epidemics based solidly on lessons learnt from past experiences, including a 'case study' in the section headed 'Detect and control' describing how PHE dealt with imported cases of MERS-CoV in 2018 and another headed 'Prepare and respond' that details the outcome of 'Exercise Cygnus' in October 2016.

While there is considerable mention in this important document of 'partners' with whom the PHE would intend to collaborate – here, among many, are mentioned, Directors of Public Health and the Local Government Association – there is no reference whatsoever to environmental health, the Chartered Institute of Environmental Health or environmental health practitioners, save for references to 'local authorities' and 'other public health agencies'.

However, as mentioned previously, the 'Coronavirus: action plan', published on 3 March 2020 (DHSC, 2021), makes reference under 'Local/Regional responsibilities' (page 22) to Local Resilience Forums and Local Health Resilience Partnerships in England, and NHS emergency planning structures in Wales, having '… primary responsibility for planning for and responding to any major emergency, including a pandemic', so these may be the 'other public health agencies' referred to in the 2020–2025 Strategy document.

Whether this answers the question why the environmental health resource available in England largely remained untapped during these crucial early weeks of the epidemic is impossible for us to say. However, what we can say from our study is how EHPs made a positive contribution where they were in post before the pandemic was declared, or when they were recruited subsequently. For instance, SM64, already a senior health protection practitioner whose work normally involved investigating complex cases and outbreaks of infectious disease, found themself managing the acute response across the whole region, as well as providing advice and support in vulnerable settings, collating data on cases and deaths and liaising with the national 'NHS Test and Trace' service.

If the circumstances at the time were far from easy – the respondent, above, commenting on the stress caused by the sheer volume and intensity of the work – a fellow EHP employed by PHE (SM41) considered it had '… highlighted the need for EHOs and their skills in providing support at a local level; skills, that were 'unsung at the moment'. However, there is just a suggestion that it may not have been that straightforward, SM41 going on to express criticism of local authority environmental health teams for being slow to respond in March and not 'loaning staff'.

4.10 Involvement of environmental health academics in the pandemic

For teaching academics, faced with impending restrictions imposed on face-to-face contact and interaction, the most pressing issue was reviewing their teaching materials ahead of moving across to remote or 'virtual' delivery. While this may have sounded a simple and straightforward exercise, the reality was very time-consuming and carried with it a form of stressall of its own.

In addition, students returning to college for the Autumn Term had a new reason to feel anxious as they encountered difficulties on campus and in halls of residence with infection spreading especially rapidly among freshers and returners in September and early October. Other students experienced challenges with IT equipment and wi-fi, while international students, unable to travel and so return to campus, were forced to study in their home country, often in quite different time zones. This placed additional pressure on academics as personal tutors performing their pastoral duties for the students enrolled on EH programmes.

While we can only really sympathise with undergraduate students of environmental health around the world for the disruption caused to their studies in the course of 2020 and now into 2021, we should like to acknowledge their patience and industry under adverse conditions circumstances, We sincerely hope that some will be recording the progress of the pandemic, the impact of control measures implemented and overlooked, and, possibly, preparing to undertake a research project along the lines of the study behind this book. Certainly, their lecturers, whether in terms of their teaching or research, have had to deal with a most unsettling and much disrupted year and, moreover, made a significant contribution along the way.

4.11 Involvement of the military environmental health cadre in the pandemic

In times of a national emergency or natural disaster we come to expect the military to be mobilised. Their capacity to deal with difficult situations, especially when these present at scale, is a source of great reassurance and deserving of our respect and gratitude.

Describing working 'flat out' trying to do normal operational activities and training in 'a COVID environment', a senior UK army officer outlined how he and a Public Health consultant had drafted a Force Health Protection Instruction document in March which had now been revised many times. This had been shared across other Government organisations and had even been used by the Football Association as the 'blueprint' in 'getting football back on track'. In addition, he was involved early on in deploying some 30 military EHPs to support civilian authorities, and developing a contact tracing and outbreak control procedure ahead of the national system for use internally and led by EH. He concluded, 'Environmental health has definitely been flying the flag – albeit in an unsung way as usual' (PC14).

Certainly, the presence of military personnel at 'Test and Trace' facilities, engaged in constructing temporary medical accommodation and in some cases supporting the police in restricting movement during 'lockdown', is a familiar sight throughout the world.

However, behind the scenes, the environmental health cadre, normally responsible for force protection and overseeing risk management and compliance with standards in military facilities, has found itself performing an important role in supporting the civilian public health response to COVID-19. However, one member of this cadre in the UK, with her routine work put on hold or else conducted by other means, was tasked with setting up, and then leading, one regional 'cell' of an internal 'test and trace' programme for the Army, while carrying out visits to military units to ensure that working practices were 'COVID-Secure' and investigating outbreaks in collaboration with the civilian Health Protection Team.

Drawing on several years in the military to equip her with the knowledge and ability to '… confidently manage outbreaks of communicable disease', the respondent (SM43) considered that the pandemic had further allowed her to demonstrate her 'leadership skills'.

One especially interesting contribution was received from a civilian EHP (PC10) working for a county council in a health and safety capacity when the authority was charged with securing additional mortuary capacity in case hospitals and undertakers found themselves overwhelmed. To avoid delay and indecision, two senior army NCOs were called in to advise the council on this, and in customary military style they did so with calm authority.

4.12 A very personal account

A former EHP, now Director of Public Health for a large municipal council, sent us her personal account of how COVID-19 changed both her, and her working life.

She begins by talking about a '… habit, nurtured years ago as a district EHO' of checking out the weekly infectious disease reports produced by Public Health England and its predecessors and so reading about a novel virus that may have made the species jump in China. She recalls talking to others about this in early January, remembering how nervous they had been about SARS and MERS, but then '… I got on with my day job'. However, from around the second week of January '… everything started to ramp up'.

The realisation that this might be the beginning of the pandemic that many had predicted, came on 31 January 2020 when the chief medical officer convened the first telephone conference with DsPH, and PHE described what was known about the virus using '… phrases that would soon become so familiar'. Many days of multi-agency emergency planning followed as cases started appearing, but by the time her authority received its first notification it was no longer newsworthy, and, as she put it, her working life had been '… completely transformed'.

Aside from the practicalities of her job, our DPH has a clear memory of the first death recorded in her borough – '… I knew his name, I wept, then we had deaths each day, I knew their names and then one day there were so many deaths I couldn't remember their names and I wept again'. She describes how the pattern of early COVID cases appearing regionally mapped to Sir Michael Marmot's report on inequalities 10 years previously, but this was just a prelude to what was to come.

Using her previous experience in dealing with major incidents, she called for the 'Gold' strategy group to be immediately convened – '… it's much better to stand these things up early' – so began her working life thereon – 12–14-hour days, 7 days a week – 'most days … a blur' – not leaving home except to get her broken laptop repaired. However, recalling driving into the council offices one Thursday morning – roads empty, being careful, 'lots of handwashing', but unable to obey the 2-metre rule – she describes her personal experience of COVID-19 thus:

> I drove home, worked three days solid and then started feeling achy and hot. I went to bed then lost my sense of smell and taste. My partner gave me a spoon of hot sauce, but I tasted nothing. I slept or laid in bed for four days, then started attending meetings from my bed ….

As the disease began to affect more and more care homes, inflicting hundreds of deaths, it also became noticeable to our DPH that communities were suffering disproportionately, notably '… our Black, Asian and Minority Ethnic communities; the communities I worked for were frightened'. In her role as 'regional emergency planning lead' she became involved in, among other things, infection control training in care homes and care-home visiting; providing public health advice to schools, workplaces, faith groups and others; setting up mobile and walk-in testing sites; writing outbreak control plans; and working on matters to do with inequality, racism and discrimination arising out of the COVID-19 pandemic.

In terms of looking after herself while working 7-day weeks, our correspondent had decided to get healthier – walk at least 10,000 steps each day and follow the 5/2 diet '… not because I wanted to lose weight but because I wanted a little control in a world with no structure'. Otherwise, she reflects on working every day from the end of January until early August, bar the days she was ill and the first anniversary of her mother's death, by which time '… you could see many of us were running on empty'. Then having stepped things down for a while, September came '… and the beginning of the second and third marathon of marathons, increasing hospitalisations and sadly more deaths'.

Reflecting on the hope for a world where vaccination, alongside physical distancing, face covering and ventilation, would provide a 'step-change' return to normality, our DPH is reminded of the need to deal with the inequalities thrown up, albeit indirectly, by the pandemic and whose '… scars will sadly be with us for a long while yet'. However, we will let her conclude the piece in her own words:

> As public health specialists we always said it wasn't 'if' but 'when' a pandemic happened. No-way did we believe it was going to be like this and I send my condolences to all those who have lost loved ones and my total respect to all who have played a part, whatever part, in managing this pandemic, thank you all.

References

Brown, P., Newton, D., Armitage, R., & Monchuk, L. (2020). *Lockdown. Rundown. Breakdown: The COVID-19 lockdown and the impact of poor-quality housing on occupants in the North of England.* The Northern Housing Consortium.

Denford, S., Morton, K., Horwood, J., De Garang, R., & Yardley, L. (2020). *Preventing within household transmission of COVID-19: Is the provision of accommodation feasible and acceptable?* Pre-print paper https://doi.org/10.1101/2020.0 8.20.20176529. Available at: https://www.medrxiv.org/content/10.1101/2020.0 8.20.20176529v2 (accessed: 24 February 2021)

Department of Health and Social Care (2021). Policy paper: *Coronavirus (COVID-19) action plan.* Available via GOV.UK at: https://www.gov.uk/government/publications/coronavirus-action-plan (accessed: 24 February 2021)

Marmot, M., et al. (2020a). *Health Equity in England: The Marmot Review 10 Years On.* London: Institute of Health Equity. Available at: http://www.instituteofhealthequity.org/resources-reports/marmot-review-10-years-on/the-marmot-review-10-years-on-full-report.pdf (accessed: 24 February 2021)

Marmot, M., Allen J., et al. (2020b). *Build Back Fairer: The COVID-19 Marmot Review.* London: Institute of Health Equity. Available at: https://www.health.org.uk/sites/default/files/upload/publications/2020/Build-back-fairer-the-COVID-19-Marmot-review.pdf (accessed: 24 February 2021)

Ministry of Housing Communities and Local Government (2021). *Guidance for local authorities: COVID-19 (Coronavirus) and the enforcement of standards in rented properties – Non-statutory guidance for local authorities on enforcing standards in rented properties during the COVID-19 outbreak.* Available via GOV.UK at: https://www.gov.uk/government/publications/covid-19-and-renting-guidance-for-landlords-tenants-and-local-authorities/guidance-for-local-authorities (accessed: 11 February 2021)

Public Health England (2019). *Infectious diseases strategy 2020–2025 – Addressing urgent threats in the 21st century* (September 2019). Available at: https://assets.publishing.service.gov.uk/government/uploads/system/uploads/attachment_data/file/831439/PHE_Infectious_Diseases_Strategy_2020-2025.pdf (accessed: 12 February 2021)

World Health Organization (2018). *Environmental Noise Guidelines for the European Region,* Copenhagen: WHO Regional Office for Europe. Available at: https://www.euro.who.int/__data/assets/pdf_file/0008/383921/noise-guidelines-eng.pdf (accessed: 24 February 2021)

5 How Environmental Health practitioners met the challenges (and discovered opportunities) arising from the COVID-19 pandemic

5.1 Introduction

So far, the focus has been on how EHPs responded to the first wave of the infection as it migrated around the world, paying particular attention to how their work and working methods changed; how their training and experience in Environmental Health served to prepare them for this moment; and how they had acquitted themselves so far, especially in terms of responding to the immediate difficulties thrown up by COVID-19, while maintaining whatever involvement was still possible in discharging their routine duties and responsibilities.

This chapter continues to tell this story, largely drawing on the survey conducted in November 2020 and summarising the responses of EHPs to questions regarding the greatest challenges and opportunities they had encountered in performing their 'new' COVID-19 work. Finally, we will represent how they thought that the contribution they and fellow EHPs had made in the name of Environmental Health might enhance the status of the Environmental Health profession in their countries, and anything they thought might still to be done to see Environmental Health properly recognised.

5.2 Early challenges encountered during the COVID-19 pandemic

There is no doubt that the pandemic has created some particularly challenging working conditions for practitioners. Managing the unprecedented volume and intensity of COVID-19 work, particularly with no end in sight (please note the respondents commented before the encouraging statements about vaccine approval and roll-out), was a source of considerable stress to many EHPs as it was to all health professionals. We present below in rough order of frequency that these obstacles and challenges were mentioned by our respondents.

DOI: 10.1201/9781003157229-5

5.2.1 Staffing, resources, and infrastructure

The impact of substantial and sustained reduction in the financing of local government, especially in terms of staffing, was frequently mentioned, here summed up by a UK EHP (SM62) in describing that their funding had been 'cut to the bone' in recent years. Others said, or suggested, that they were just 'spread too thin' to maintain both their COVID-19 work alongside other vital EH work, the latter suffering badly or sometimes suspended altogether, an issue described in detail in Chapter 3. SM18, from South Africa, was also concerned that resources would continue to be constrained because central government had exhausted its 'COVID relief' funds.

As far as local government was concerned, EHPs felt that they were doing the best they could, but their COVID-19 work was neither as responsive nor dynamic as it should and could have been. Some were unable to complete their work in the time available so had to resort to overtime or putting routine work aside. There was, then, the on-going challenge of developing and then scaling up COVID-19 responses at speed, at just the time that the Environmental Health profession was, according to SM27 in Scotland, '... confronting a staffing crisis with an aging workforce and low recruitment'.

Across local governments, EHPs described shortages in staffing, both historically and through COVID-19, where the member of staff fell ill, or a dependent household contact had to self-isolate. The pressure to work overtime, usually unpaid, was a constant source and concern for some, with an American (SM39) admitting that they were continually working 100-hour weeks, while seeing first-hand the physical and mental strain that the infection was having on all staff. For them, watching their colleagues 'break' was the greatest challenge of the pandemic so far.

Other obstacles to being able to make a more effective contribution included working with antiquated IT, and when working at home, lacking IT support. In addition, the lack of administrative support was hampering others in their pandemic work, though some felt they lacked wider support in areas like mental health and well-being.

Based on the survey alone, most EHPs continued to work out in the community with its many challenges, not least to their own health. Clearly illness from COVID-19 has affected EHPs in the field, a South African respondent describing how much their services had been affected by this. The increased risk arising from shortages of PPE for EHPs, notably in South Africa and Zambia, in turn compromised their ability to promote 'good prevention' measures in the community. The damaging impact of the virus was brought home to a UK EHP (SM44) who reported that their manager had resigned because of COVID-19 work pressures, and SM52 (country unknown) admitting, '... we have fallen apart as a department'.

Wider resource problems included a lack of health education materials and transport in South Africa, while SM24, in Nigeria, described a lack of transport and other basic tools for their COVID-19 response work including mobile phones, measuring tapes and training in legal procedures. Furthermore, lack of transport, particularly during periods of lockdown, made local level case finding and contact tracing far more difficult.

5.2.2 Keeping up-to-date and making sense of legislation and guidance

Navigating the uncertainties of COVID-19 policy, particularly legislation and guidance, was a common challenge reported by EHPs and contributed considerably to their heightened level of stress. Constant policy changes were the norm, often at very short notice, and on certain occasions, received only the night before implementation. EHPs said that they were especially unhappy to learn of changes via the media rather than the government itself.

Beyond this, EHPs reported that policy documents were sometimes poorly drafted and the meaning unclear, with governments slow to clarify how terms should be interpreted, so their meaning remained vague. Furthermore, from time to-time, COVID-19 guidance did not match up with legislation, or one set of guidance conflicted with another, and it was not uncommon for guidance to be published some time before the legislation it was intended to support, or a long time after.

Not surprisingly, very few were able to comment on the situation in two or more countries regarding the same issue, though one, a private sector EHP (SM37), considered the Scottish Government's website was 'cluttered and very difficult to use', though they considered it was noticeably better at coordinating guidelines than their English counterparts.

The issue of inconsistency of interpretation between local authorities in the same country was of concern to many EHPs since this led to uncertainty and gave businesses and individuals the opportunity to exploit 'loopholes'. However, in the words of an EHP working for a Dutch national government agency (SM14), it provided '... a gravy train by which to profit'.

More important to public health, several EHPs considered that inconsistency, and even more so uncertainty, could prejudice the outcome of efforts to limit the onward transmission of the virus, with a South African EHP working in local government (SM35) commenting that during periods of high community transmission, uncertainties about modes of transmission made it even harder to advise businesses and communities on control measures.

Such was the concern regarding policy uncertainties that one UK EHP (SM44) summed up the, then, new COVID-19 laws and powers as 'rubbish', suggesting that some vital health protection powers simply could not be used. However, an American EHP working for an NGO (SM15) believed that, in the United States, the interface between science and policy had failed spectacularly over COVID-19 and, unfortunate to say, the Environmental Health professional was partly to blame for this.

By way of illustrating how important it is to frame legislation very carefully and with due regard to any potential conflict with local cultural beliefs, a South African local authority EHP (SM13) described the upset and animosity among the families of the bereaved and the wide community when the instruction was that the body of the deceased should not enter the house at a funeral.

5.2.3 Not being recognised or appreciated

Thinking back to Chapter 3, when we considered preparedness to respond to the disease at the outset, while most EHPs thought their education, training and experience prepared them well for COVID-19 work, this was not necessarily recognised by their employers and therefore served as an on-going obstacle. A UK EHP in local government (SM42) described being 'ready to go' and full of ideas at the start of the pandemic but there was concern expressed that colleagues from outside Environmental Health might have felt intimidated.

Other local authority EHPs reflected on doing a good job on COVID-19, but in their opinion, other decision-makers – citing politicians and senior managers – failed to fully understand what Environmental Health was about and therefore EHPs were not consulted on local COVID-19 matters. Expressing their view that this was regrettable, and though working with politicians was itself a challenge, a Channel Islands EHP (SM59) pointed out that briefing them on what were sometimes unpalatable truths represented a 'vital role'. However, a local government EHP (SM60) from the United States expressed frustration when their governor's own communications contradicted national COVID-19 guidelines, thus adding ambiguity to an already confused situation.

5.2.4 Difficultly in establishing channels of communication and understanding

The value of working across different tiers of government was clearly recognised by our respondents as creating many of the opportunities described below, though not without challenge. In Australia, a regional government official (SM40) commented that while COVID-19 issues required a rapid and 'whole of government' response, this was not always possible due to 'bureaucracy'. On the other hand, in South Africa, local authority EHP (SM17) bemoaned the absence of a national reporting system for COVID-19 which they put down to poor communications between local, provincial, and national government.

A more significant problem was caused at State level in Australia when, according to a local authority EHP (SM50), having made the police responsible for COVID-19 law enforcement they were reluctant to enforce the law in food premises because this was outside their area of expertise. More fundamental still was the situation in Liberia, about which SM31, speaking on behalf of EHPs working at a local level, commented, that, despite being fully prepared and ready to respond, the national government was '... not up to the task', to the extent of not being able to pay EHPs in a timely manner.

According to a local authority EHP in Scotland (SM27), a degree of mistrust existed between central and local government because of a reluctance to release further COVID-19 resources for their work, preferring instead to invest in private sector partners to deliver its COVID-19 response, particularly in areas like contact tracing. This was not helped, they felt, by the impending end to the Brexit transition period on 31 December 2020.

It was interesting to note that the EHPs surveyed reported significantly fewer 'challenges' arising from their routine 'day-to-day' work beyond those related to the shortage of resources (mainly human) and uncertainty over policy described above. The one exception was a UK private sector EHP (SM61) who was receiving all the calls from tenants on disrepair matters that should have been picked up by the Council's private sector housing teams.

5.2.5 Intransigence and denial

Although there was general concern about the refusal to co-operate with local authorities, respondents in South Africa reported that many businesses and members of the public were not taking the pandemic seriously, or that businesses and communities would make improvements but then not maintain them. Sometimes this was because legislation conflicted with cultural norms, as when the funeral arrangements described above caused offence, but another South African EHP (SM35) spelt out the tensions between COVID-19 rules and local realities, when they commented,

... socio-economic factors in our communities made things hard for us ... for example, isolation of infected individuals was impractical in certain areas due to the fact that there's not enough rooms for that in a household.

Although a disturbing and troubling phenomenon world-wide, African EHPs, in particular, singled out the spread of 'fake' COVID-19 news creating constant confusion and uncertainty across communities. In Nigeria, EHP (SM24) reported that, as a consequence, people were often reluctant to come forward for COVID-19 testing, while in Zambia, a private sector EHP (SM58) was concerned that even some of their own team members thought COVD-19 was a hoax.

5.3 Opportunities presented by the COVID-19 pandemic

Having previously established that EHPs felt they had the necessary suite of skills and abilities to respond to the appearance of COVID-19 in their countries, we hoped to establish whether they had found an opportunity to use them, on what scale, and to what effect.

5.3.1 'Making a difference'

Despite the many challenges mentioned above, when asked what the greatest opportunities had been so far, EHPs were generally proud of their many different contributions to the public's health, with one, working for a local regional agency in the UK (SM64), speaking of feeling 'part of history' and another (PC2, also UK) speculating as to the number of lives that he might have been responsible for saving during the pandemic, reflecting on the impact of the health protection role performed by EHPs under normal circumstances, and concluding that EHPs were '...overlooked and undervalued'.

The sense that through their actions they were making a difference to people's lives came out strongly, especially, in the case of a local government EHP in South Africa (SM17) commenting on having 'closer contact with the communities' they serve. For an EHP in Portugal (SM33), there was a suggestion that prior to the arrival of the pandemic, Environmental Health had 'stagnated' and held little value in Portugal, but this had changed now, that they had had an opportunity '[To] work in the public health area, exercising my degree'.

For an EHP in the UK (SM56), the opportunity to make a difference came when they were called to '... help business to stay trading by assisting in advice on controls', while for another in the UK (SM57) it was '... being able to work with businesses to manage outbreaks of COVID and to do a job where you can see a direct impact of your actions'. For some – here quoting again an EHP attached to a region of a national agency (SM64) – '[it] has highlighted the need for EHPs and their skills in providing support at a local level, [they] are unsung at the moment'.

5.3.2 Strengthening relationships, both internally and externally

For some, COVID-19 had strengthened relationships within their own teams, for example, a local authority EHP in the UK (SM42) spoke of Environmental Health, once again, 'rising to the occasion', but going further and indicating that their technical and administrative staff had all, '... stepped up, taking calls, shopping, chatting, engaging, enforcing. It's been, for me, a real team effort'.

Others commented how relationships between EHPs had also been strengthened nationwide, helped by initiatives like the COVID-19 webinars offered by the CIEH in the UK and similarly in South Africa. Noting that tensions remained with local communities and businesses, local government EHPs widely valued the opportunity for greater engagement with them during the pandemic, particularly in the provision of public health support and wider awareness raising activities.

Reflecting that COVID-19 would be with us forever, a South African EHP (SM9) opined that recent opportunities to build community relations were a good investment for the future. Meanwhile, SM37 thought the creation of 'COVID certification standards' had helped to raise compliance among their clients and demonstrate the capabilities of staff. Otherwise, a private sector EHP (SM19) saw benefits accruing through greater engagement with their clients, saying also that the pandemic had raised the profile of environmental health among senior managers.

5.3.3 Building professional partnerships through 'joint working'

Mention before has been made to the value reported by EHPs of joint working with other teams, particularly those in Public Health, bringing regions together and cutting across different tiers of government. These included learning new methods of pandemic response through networking with specialists in communicable disease, and securing access to their advice, training, and communications resources, along with specialist databases.

For a Channel Island's EHP working for the State government (SM3), the pandemic had created an opportunity to raise awareness of environmental health services among ... our public health colleagues. Their Channel Island colleague (SM6) was at pains to attribute the success of the operation to the leadership qualities of the Director of Public Health and the support of staff working across their organisation. Similarly, for a UK EHP working in local government (SM5), this was an opportunity 'to shine professionally' and get involved with EHP colleagues and other professionals at a regional and national level.

Some especially positive outcomes emerged from the United States, where a county agency EHP (SM20) spoke of new partnerships with other agencies and business itself coming out of the pandemic. In future these might help in managing, if not preventing, local outbreaks, but had already enabled EHPs, particularly newer staff, to gain experience and develop new skill-sets. While this was particularly heartening to hear, it was a little disappointing that it took a global pandemic to bring about partnership working in many parts of the world, with several respondents saying that they thought that without the pandemic they would never have met colleagues in areas like Public Health or Trading Standards and they were hopeful that these partnerships would continue after the pandemic.

5.3.4 'Joint working' – not new ... and not always without its challenges

Indicating that joint working was not a new phenomenon in some countries, an experienced academic and former EHP based in Wales (SM11) was able to speak to the fact that 'collaborative networks', were already well established in Wales before COVID so that the Environmental Health profession was '... at the heart of the public health response', particularly assisted by the 'Lead Officer in Communicable Disease Control' system in place, which has paid dividends. This, and having experienced EHPs in every Welsh local authority, meant that they were '... immediately able to step up and take on important COVID-19 related roles'. This may account for the fact that some were now '... co-ordinating regional contact tracing teams', while others were 'leading local authority responses to Covid-19', with lead officers having regular contact with Public Health Wales.

However, against this positive backdrop, we heard that 'joint working' was not all plain sailing. Reflecting on his work on COVID-19, one

UK EHP employed in the private sector (SM22) recalled a degree of tension with Public Health practitioners, something he thought came down to 'professional snobbery'. Another, this time working for the state government in Nigeria (SM24), while applauding the new-found opportunity '... to work effectively in teams of diverse professionals', reported team leaders from a non-Environmental Health background, sometimes undermining their EHP staff in areas where the latter had expertise.

5.3.5 Teaching environmental health – finding 'opportunities' under adversity

We have heard already how difficult and stressful this was for the staff, especially when they were under pressure to perform COVID-related activities within their institutions. Here, one university-based EHP (SM11) describes the need for entire programmes in the 2020/21 academic session to be '... transitioned into flexible, blended learning with short bursts of on-campus teaching plus extensive online and virtual classroom learning (using Moodle and Microsoft Teams)'.

Yet, putting this aside, an Australian academic (SM49) commented on how moving from classroom teaching to largely online had equalised access to resources for students, thus, potentially, allowing more to benefit. In addition, the transition had prompted the need to improve the quality of online resources, including the use of practical simulations, and a UK EHP lecturer (SM21) spoke of the opportunity to appear regularly on regional and national radio and television to discuss COVID-19.

5.3.6 Opportunity to ensure job security and enhance career progression

The precarious nature of employment during a protracted international crisis has been apparent, that even when the service provided is deemed 'essential', and the skills of the deliverers are in great demand, jobs are at risk. Clearly, there are forces at play throughout the world to cause practitioners to feel uncertain for their immediate future, let alone a future without pandemic. Here, the difference between richer and poorer countries was apparent, with two EHPs from the former indicating that, having had the opportunity to demonstrate their abilities and leadership skills, they were now working at a higher level than before the pandemic, and a UK EHP (SM42) believing it would look good on their CV!

Jerry Chaka, in his report outlined in Chapter 3, saw the pandemic providing opportunities and benefits for the profession in securing employment for previously unemployed EHPs, and in South Africa seeing '… most of the newly-qualified Community Service Environmental Health Practitioners placed by the Department of Health at different municipalities where the need was high to assist with the COVID-19 pandemic'. In so doing they would be able to complete their one year of community service required for registration with the Health Professions Council of South Africa and therefore eligible to practise independently anywhere in the Republic of South Africa. On pay, Zimbabwe EHPs are now entitled to receive a 'COVID – Risk Allowance' which has improved their remuneration, while some of the municipalities in South Africa are paying their EHPs 'danger allowances' due to the COVID-19 risks.

However, respondents to our survey, one, a local government EHP believed to be from Africa (SM52), said that they grateful to keep their job when so many were being made redundant or relying on government support to pay their wages. A Nigerian EHP (SM16) was clearly looking ahead when he said that he hoped to be recruited soon by an international health agency, while others thought that their COVID-19 experiences would be good for career advancement.

5.4 How the COVID-19 pandemic altered perceptions of Environmental Health

5.4.1 'It has altered perceptions of EH for the good'

Nearly half of the survey respondents indicated that they thought it had, though most mentioned the many and various challenges they had faced and had to overcome. Interestingly, some talked about this having played a part in self-discovery of their worth, for example an EHP at a university in Slovenia (SM10) saying they have '… rediscovered the importance of our profession', and several indicating how and why this might have come about, for example an EHP working for an NGO in United States (SM47) suggesting that it was because Environmental Health had had to work '… outside of its normal bubble'.

Several believed that it had already achieved visibility for the profession, mentioning that the minister of health in South Africa had finally recognised '… the profession and the role we play in communicable disease control' (SM17), and the government, who will '… hopefully offer more support and resources' (SM53 – local government EHP in South Africa).

However, 'Hope' seems to have been an emotion in strong supply among EHPs at this time, with a local government EHP in Portugal (SM33) responding,

> I hope that we are seen more in society as a profession with great importance in public health, as a profession of value, with great professionals. I hope that we have appreciation and respect, and that after COVID-19, we will not be forgotten again.

5.4.2 'It has altered perceptions of EH ... but more needs to be done'

For a quarter of respondents there was the suggestion that more needed to be done to sustain and build on the many achievements thus far. This meant, for a university lecturer in the UK (SM21), continuing to demonstrate to others – governments and society – the importance of Public Health and Environmental Health, thus maintaining it in 'public consciousness' (South Africa local government EHP35). A fellow South African EHP (SM35) saw this moment as the '... greatest opportunity to influence decision makers to take this field, prevention, very seriously', observing that '... Environmental health has never been spoken of until this pandemic hit', and opining that their government now realised the very considerable role played by Environmental Health and EHPs in preventing people from becoming infected, referring to Environmental Health as '... the forgotten profession', UK local government EHP (SM55) thought that the skills of EHPs had '... shone in the pandemic' and that with the correct handling, this was an opportunity '... to rekindle our profession'. While concurring with this sentiment, another UK EHP (SM64), this one working for a regional agency, made this telling comment:

> I have seen some amazing work by EHPs during the pandemic. They have stepped up and gone beyond what's expected. However, they need to be championed as they tend to be so humble and quiet!

Agreeing that the pandemic had '... raised the role of public health and environmental health', UK local government EHP SM27 considered that this had come at '... exactly the time when the EH profession is confronting a staffing crisis'. A private sector EHP in Germany, SM28, thought that the pandemic had drawn attention '... to the importance of the work carried out by EHOs and other similar competent health professionals', though calling for more targeted funding to make these '... operational resources readily available'.

5.4.3 'It has altered perceptions of EH ... but far more needs to be done'

Some respondents made the point that rather than just national governments seeing (or needing to see) Environmental Health in a fresh light, so did employers and government agencies, with EHPs in the UK believing that local government and agencies like Public Health England should value their environmental health workforces more. Several went on to suggest that after years of cost cutting and under-funding of services, the consequences, as one Dutch EHP (SM14) put it, were 'biting back very hard now', suggesting the need to invest in the recruitment and training of EHPs.

The shortage of EHOs in all states of Australia is clearly an issue, which a spokesperson suggested was down to '... a lack of understanding of the environmental health role' especially in the context of public health. Similarly, while excited at the prospect of professional opportunities opening up in the future in Australia, SM40 thought that some states/territories would be more likely to invest in their Environmental Health workforce than others.

A rather hopeless conclusion was reached by a South African practitioner-lecturer (SM12) in respect to whether there was a means of sustaining whatever good might have come during 'lockdown', when they observe that '... the strengthening of Environmental Health during COVID-19 will only last for the ... period of the crisis', it being the nature of such things going back to the 19th century, and that unless the opportunities provided by the pandemic were 'optimised', what advantage might have been achieved would normally '... after a while, slip back', and no longer '... be valued by decision-makers'.

Clearly others agreed with this sentiment, though were prepared to offer suggestions as to the means of building on Environmental Health's enhanced status, with SM22 calling on the professional bodies to publicise the work performed by EHPs and '... getting in the ear more' of government departments and agencies, and SM15 in the United States indicating the need to 'showcase' the essential roles performed by EHPs, share lessons learned globally, and revisit the 'marketing and branding' of the profession.

Arguing that they thought the environmental health challenges exposed by COVID-19 were universal, a South African university lecturer, SM34, called for a 'global professional strategy', the absence of which, a

Zambian EHP (SM58) maintained, was a matter of grave concern given that COVID-19 had exposed the fragility of the country's economy and health systems.

5.4.4 'It hasn't altered perceptions of EH yet ... but it still might'

The remaining one-fifth or so of respondents were uncertain whether the response to the pandemic would strengthen the profession's position or not. On the one hand it had raised the profile of Environmental Health, with a South African university EHP (SM12) suggesting it had become 'everyone's business', while others thought it still lacked 'visibility' and its practitioners remained 'unsung', with a UK EHP – SM42 – questioning whether governments would deliver on commitments made during the pandemic to invest in Environmental Health services.

Others drew attention to EHPs finding themselves relegated to performing minor roles, with a South African EHP (SM48) referring to being assigned 'trace and trace' when other professions led on proactive work. Similarly, a Liberian EHP (SM31) was concerned that EHPs were still not leading on outbreak control work. In the UK, where many professions were involved in COVID work, SM62 considered that EHPs remained 'last in the pecking order' with insufficient resources to undertake their pandemic work. Blaming the slow response to COVID-19 by the environmental health profession in the UK, thus finding itself initially being side-lined in favour of Public Health, SM41 paid tribute to the EHPs on temporary contracts and their significant contributions subsequently.

Agreeing on the point about the slow response of local authorities in the UK, which in some cases came down to lack of support, a private sector EHP (SM37) also thought very poorly of local authority EHPs '... posting on social media how excited they are to serve prohibition notices on businesses who are not implementing COVID standards' since, in their view, it would damage respect held by hospitality providers of those delivering environmental health services in their areas. They go on to suggest that this is also not in the spirit of 'everybody being in it together' which the respondent had not witnessed during the lockdown, something that might impede efforts to strengthen the profession in years to come.

Similarly, SM19, another UK private sector EHP thought that he EH profession should have spoken up more for the hospitality industry given that '... government insists that this is the root problem' and knowing the amount of money spent '... to ensure compliance and create safe environments'. Pointing out how UK Hospitality had been 'a loud voice' on the subject, they conclude that this has been '… an area CIEH have missed/ignored'.

However, few of the respondents saw the situation as irretrievable. Even one of those remarking critically on the behaviour of enforcement officers in posting on social media above (SM37) felt that '… the proactive and helpful approaches by enforcement authorities will contribute to strengthening the position of our profession moving forward'.

Just five of those responding to this question felt here was little or no hope, though one must remember that this survey was conducted in November, thus before the full extent of the challenges of the second wave were apparent. Of those explaining why they held this opinion, one (SM44), reflecting on all the good work EHPs had done, thought that other professions had drawn the plaudits, and that environmental health was still not recognised, consulted, or publicised, further commenting that despite the need to recruit more EHPs, environmental health qualifications were being watered down and instead '… EHPs were fast becoming a dying breed'.

Another (SM56) discounted the lack of publicity, reflecting that despite the (then) recent coverage, it pointed in their opinion to central government still not really understanding what EHPs did, while SM46 asked what might happen to all the new COVID-19-related jobs once the pandemic ended. An American local government EHP (SP60) was also sceptical that things would improve for the profession, for despite attention being drawn to the importance of Public Health professionals, local government continued to dwell on the past and cut staff.

A rather bleaker view was offered by UK local government EHP5 who thought that wider agendas were at play and considered the UK Government was using COVID-19 as,

> … an opportunity to break local government financing on a national scale as a pre-cursor for the reorganisation/centralisation of governance and services in the UK for dogmatic reasons, rather than for the benefit of the population. This can only be to the disbenefit of the profession.

A final comment from a local government EHP for the United States (SM39) sums up the conundrum that is Environmental Health:

> This profession is not glamorous; you are only quietly rewarded for your efforts. No one knows the work you do, which is why a lot of the people in this profession do it. We do not want to be the face of movement, just the gears and minds to be out there protecting the public…

6 Support for the practitioners

6.1 Introduction

The 'support' discussed in this chapter is that to which practitioners might turn when looking to see, for instance, how they were being represented and promoted; whether they were thinking and working along the right lines; whether others had experienced similar problems and found a novel solution that they might emulate; and when, on a personal level, reading about how others were feeling and coping. At times of impending crisis, especially when the problem itself might be uncertain, let alone the solution, practitioners might be expected to look to many other organisations for advice and professional guidance.

It would be an interesting exercise to discover where practitioners have sought support in the first nine months of the COVID-19 pandemic. We conjectured that this might include non-governmental organisations (NGOs), professional bodies and membership organisations, trade bodies, technical and representative groups, and academic institutions. In addition, we knew that practitioners had joined ad hoc groups through social media to share their experiences and obtain support.

As we know, COVID-19 did not strike every country simultaneously and with the same ferocity. However, as mentioned previously, 30 January 2020 stands out as the moment in the disease's epidemic history when the world at large was alerted to its public health importance through the director-general of the WHO's declaration that the novel coronavirus outbreak was a public health emergency of international concern. Allowing time to assemble material, and putting this into a form suitable for publication and securing agreement from the organisation's governance to release it, one might have expected there to be supportive advice and guidance available to practitioners by roughly mid-February.

Setting aside the many organisations that have chosen to produce material to better inform the public on the virus, disease, and pandemic, here we mention some of those that looked to support EHPs. To an extent,

DOI: 10.1201/9781003157229-6

the COVID-19 pandemic has been a 'test' of the capabilities of these professional and membership bodies to offer their members timely and relevant advice and guidance on matters pertinent to the virus and its control. Judgement as to its usefulness or otherwise resides with the EHPs who may have consulted these organisations and accessed the material. Only where the efforts of these bodies have, in the view of those that responded to us, not met their expectations have we drawn attention to this, and questioned why this might have been the case.

6.2 Professional bodies

Although by no means all EHPs of the world have an opportunity to draw support from a professional or membership body in their country, those that can access material from their own and other organisations without difficulty tend not to find this behind a 'pay-wall'. Furthermore, the research titles featured in this book have, in the most part, expressly decreed that all material pertaining to COVID-19 should be made available on an 'open access' basis.

However, before describing what the professional bodies representing EHPs around the world were saying and doing about COVID-19, it seems appropriate to start with the International Federation of Environmental Health (IFEH), the organisation created as a global confederation of the various professional bodies, and describing itself on its website (https://ifeh.org/about.html) as '… a community of environment health professionals and academics, working together to disseminate knowledge concerning environmental health and promoting and supporting co-operation on environmental health issues worldwide'.

6.2.1 *International Federation of Environmental Health*

The COVID-19 emergency put the International Federation of Environmental Health to the test of meeting the objectives of a new Strategic Plan that its Board set the incoming president to implement during his term of office. Reflecting on this in his 'Past-President's Report – 2018 to 2020', Dr Selva Mudalay believed that he had delivered on his remit (Mudalay, 2020) and one would have to say that it was already on the way to achieving its 'Vision' of being '… the internationally recognized leading professional organization in the field of Environmental Health' and its 'Mission' that begins, 'To promote international co-operation between Environmental Health professionals …'.

As evidence of the latter, we observed that the IFEH was quick to offer guidance to all who might have resorted to its website, posting a piece entitled 'Coronavirus – COVID-19' on 8 March 2020 (IFEH, 2020a). Apart from commenting on the emerging situation and providing, then, rudimentary advice on infection control, it described the role and common goal of Environmental Health. This is telling in that it makes no reference to the regulatory role, preferring to say, 'Environmental Health Practitioners have a wealth of expertise in multiple disciplines such as epidemiology, toxicology, food safety, waste management, among others.'

Writing in the Federation's December 2020 Newsletter (IFEH, 2020b), the incoming president, Professor Susana Paixão, wrote, 'The SARS-Cov-2 pandemic has come to demonstrate the vital role of the Environmental Health workforce worldwide to face the challenging time for all humanity', adding further, 'I hope that we will be able to take advantage of the window of opportunity to raise awareness of Environmental Health …'.

Choosing to act as a conduit for the contribution that EH and EHPs had made in member countries in the first months of the COVID-19 emergency, the IFEH posted updates mid-year from those representing the discipline and its practice entitled 'COVID-19 Pandemic Reports'. Material from these and other national update reports have been used elsewhere in this book, and aside from being listed and referenced at the close of this chapter, we wish to acknowledge and thank the respective correspondents for these insights, and the IFEH for making these available to all.

6.2.2 Environmental Health Australia

The EHA seems to have taken up the cudgels on COVID-19 in late March with communications to its members informing them of the current situation and the resources available. However, an Open Letter from National President, Philip Swain, to Ministers dated 22 March 2020 (Swain, 2020), demonstrates the EHA's preparedness to put itself forward as the voice-piece for the EH profession in Australia, and willingness to offer its opinions, on this occasion believing that 'the current measures that have been introduced are insufficient to check the progress of the disease'. The letter goes on to recommend '…more extensive tracing, testing and self-isolating of the contacts of all confirmed cases, enlisting the assistance of Local Government EHOs to assist with the work', and calling for clearer advice to be issued on self-isolation and social distancing through 'saturation of traditional and social media' outlets.

In a further communication on 9 April 2020 entitled: 'EHOs and their role in the community in response to COVID-19' (EHA, 2020a) the EHA describes its membership base of Environmental Health Officers as 'experienced and expertly qualified, making clear that EHOs, with their support '...will continue to support Public Health Units and local authorities', playing '...an integral role in the national response to COVID-19.

Reviewing how State and local government Environmental Health providers had responded to the pandemic up to 24 May 2020, some four months after the first case was reported in Australia on 25 January 2020, the EHA's COVID-19 Pandemic Report is a source of data that has been used previously (see Chapter 3). A feature of this Report is an Appendix that contains a series of COVID-19-related infographics that the EHA says it put together '... independently and in partnership with ECU, and EHA accredited university' (EHA, 2020).

6.2.3 *National Environmental Health Association (the United States)*

Formed in 1937 and now boasting a membership of 6,000, NEHA represents registered Environmental Health specialists or registered sanitarians working in EH in different capacities around the United States. NEHA's Mission Statement, 'to advance the environmental health professional for the purpose of providing a healthful environment for all', is described on its website as being delivered through 'programs' that are '... both an educational as well as a motivational opportunity'.

As far as COVID-19 is concerned, NEHA announced its intention to commit to its 'mission' of supporting practitioners through a dedicated 'COVID-19 Pandemic Information and Resources Webpage' released on 15 March 2020 (https://www.neha.org/covid-19). One gets a strong sense from reading the chief executive officer Dr David T Dyjack's, introduction on the front page of the 'NEHA – COVID-19' section on the website (https://emergency-neha.org/covid19/), as posted on the IFEH website on 19 March 2020, that NEHA was expecting EHPs to 'step up to the plate' when it came to COVID-19. Here, he seeks not so much to motivate as to exalt members to make their mark during the weeks ahead, when, as he puts it, '... like Atlas of Greek mythology, environmental health professionals will be asked to hoist many burdens upon their shoulders', believing that '... the stress promises to be intense and relentless' and the issues will be 'emotional and compelling'.

Seeing Environmental Health professionals across the United States and around the world working '... on the frontlines of preventive public health services delivery', he commits NEHA '... to supporting the

EH workforce to effectively and safely do their jobs', especially important since Environmental Health professionals in the United States form the second largest part of the local Public Health workforce, after nursing.

Picking this up again in his 'Open Letter to the Environmental Health Profession' on 9 April 2020, Dr Dyjack observes,

> This is the moment for which we were trained. Let's bring a contemporary interpretation to the art of practice. Let our face be one of kindness and empathy. Let our science drive our community's recovery and resiliency. Let this be the moment where we safely shepherd the frightened people into the future.

He finishes,

> Let's create a profoundly memorable moment, one that undeniably demonstrates the value of our profession to the health and prosperity of the communities we serve.
>
> Dyjack (2020)

Of the various practical initiatives performed by the NEHA, the Environmental Health workforce assessment surveys stand out as a model for other countries. NEHA have been quick to see the importance of identifying how their members were putting their skills and experience to good use and, conversely, how COVID-19 was having an impact on the workforce. They set out to discover this through a survey distributed to members by various means on 25 March 2020, thereby obtaining a snapshot of their 'actions and needs' early on.

Appearing in a report released on 1 April 2020 entitled 'COVID-19: Environmental Health Workforce Rapid Needs Assessment Report' (NEHA, 2020a) it represents the responses of some 1,175 'local, territorial, state, tribal, federal and private sector respondents'. The survey provides valuable data, some of which were used in Chapter 3. However, among the 'additional concerns and issues' being faced by survey respondents was this one:

> It would be great if NEHA also addressed things like burn-out and stress. This is stressful for everyone, but especially stressful dealing with people who are having to shutter their businesses or take precautionary steps that they don't feel are warranted [but] thanks for asking ….

Subsequently, NEHA undertook a more detailed survey of the Environmental Health workforce over a six-week period (15 July 2020–31 August 2020) some four months after the pandemic was declared, seeking responses from EH professionals working in both the public and private sector and at federal, state, and local level, and others working in tribal and territorial government. Seven hundred and sixty-seven responses were received, over half from 'local EH programs', the aim being to '… assess EH workforce activities and identify needs in response to COVID-19'. The findings are provided in a report published in October 2020 entitled 'COVID-19 EH Workforce Needs Assessment Report II' (NEHA, 2020b).

It is suggested that 'Report II' provides a model as to how other IFEH member organisations might perform similar assessments of EH activity and practitioner 'needs' in their countries or across states and regions. It is especially praiseworthy as it purposefully addresses some of the key shortcomings discovered in the March survey, and rather than shying away from difficult topics, explores the twin issue of burn-out and stress in detail, accepting that this might invite criticism of employers and NEHA itself.

Thus, against what the authors of the report regard as 'unequivocal' – that EH practitioners are '… actively supporting COVID-19 response and recovery' – this is coming at a cost as they take on new roles and responsibilities, some of which bring them into harm's way. When asked about their inner capacity to cope with this, nearly three-quarters of respondents 'agreed strongly' that they felt 'emotionally exhausted' by it, citing, repeatedly, the challenge of having to balance the demands of their everyday work with COVID duties, or as one practitioner put it, '… feeling the heavy weight of competing priorities'.

Another responded on the issue of stress caused by the pandemic in a way that would likely chime with EHPs around the world, similarly working '… consistently long days and [at] weekends', and saying,

> The stress of the pandemic in general, coupled with the sudden increase in work, lack of guidance, and the enormous pressure of trying to manage a pandemic while dealing with seemingly endless angry and upset people has been incredibly difficult and taxing.

6.2.4 Chartered Institute of Environmental Health

With a membership of around 7,000 qualified practitioners and associates, and a history going back to the dawn of the Public Health movement, one might have expected the UK-based CIEH to have been quick to make a statement or offer support for members about the fast-developing epidemic in February. Yet the first tangible evidence of support for members working in the field (though also directed at 'the wider public') seems to have come on 6 March 2020 with a press release posted on the website entitled 'CIEH issues coronavirus guidance' (CIEH, 2020a). This included a short video on handwashing and a link to the UK Coronavirus Action Plan, issued by the Government on 3 March 2020 and discussed previously.

It is understandable, though unfortunate, that this official document may have caused the writer of the piece to accept the premise that since coronavirus was, in his words, '... fundamentally a public health issue and the appropriate agencies for advice and guidance to the general public, businesses and organisations are the public health agencies and Health Departments of the United Kingdom', that it would not be necessary to assert the case for EH/EHP involvement. However, it still came as a surprise to read:

> Although Environmental Health Practitioners (EHPs) do not have a lead role in tackling coronavirus, they do provide support to the relevant government agencies and organisations and will potentially play a role in local authority emergency plans for Infectious Disease, should these plans be actioned.

In what may have been prophetic, the CIEH seems to have accepted the inevitability of the transition from the 'Contain' to the 'Delay' phase in England, Wales and Northern Ireland, suggesting that it was important 'to take a balanced and reasoned approach to combat the spread of coronavirus and follow and keep up to date with government advice as we start to progress from the Containment phase to the Delay phase', while indicating that as the situation develops, the CIEH would issue further guidance 'if necessary'.

A rather different and far more positive line was communicated to members via the CIEH online platform 'EHN Extra' ('The CIEH news bulletin') sent by e-mail by its editor, Katie Coyne, on 19 March 2020 referring to the country heading into '... one of its greatest public health challenges in history' and drawing attention to the '... the team at CIEH ... working to support, inform and encourage [them]'. Now, the message is clear, EHPs are '... uniquely placed to help with this crisis', later urging Government to put EHPs '... on the front line against COVID-19'.

In an appended article from 'The CIEH Team' entitled 'CIEH: supporting you in the fight against coronavirus' (CIEH, 2020b), it is confirmed that 'EH teams' are already '... at the frontline of mitigating the epidemic' while recognising the need to continue '... keeping the public safe from other hazards' and describing many members as already being 'overstretched'. It goes on to indicate what the CIEH itself was doing, the list including: summarising material coming out of government, 'gathering and sharing good practice'; 'compiling a region-by-region list of people who have contacted us offering their services as volunteers'; and contacting local authorities, public health bodies and others '... offering volunteer capacity as needed'.

On this final point, Katie Coyne writes in the same communication of EHPs being ready to meet the challenge of the COVID-19 crisis, and of members expressing their concern that they were being 'overlooked as an important resource' and local authorities 'complaining that the government's plans may not appreciate the extent to which they can help', sentiments that were sometimes expressed by respondents to our survey. Under the heading 'Use us to fight the virus, It's our job' (CIEH, 2020c), Coyne, quotes, among several others, an EHP at Westminster City Council, much frustrated at seemingly being passed over despite having the skills and local knowledge to assist. As the young EHP put it, 'It's like EH is being told to "sit out" ... why are we being left out when we are eager and keen to help? I shouldn't be sitting on my hands'.

As far as technical and professional support for practitioners, from the first day of 'lockdown' in the UK (23 March 2020) the CIEH became a prolific source of material for practitioners, whether through its online platforms, 'EHN Extra' and 'Member Connect', personal e-mail communications from the Executive and directors responsible for Wales and Northern Ireland, postings on the CIEH website and articles in its hard-copy magazine for members 'Environmental Health News' (EHN) from April. Subsequently, material has been posted on a 'Resources' page (https://www.cieh.org/policy/coronavirus-covid-19/), while two guidelines drawn up internally, but with assistance from others, were duly shared with all IFEH member organisations and relevant EH professionals through the IFEH website.

One innovative way that CIEH has sought to engage with members and, interestingly, non-members, is through a regular series of webinars called 'COVID-Conversations' that have variously featured the two CIEH Directors for Wales and Northern Ireland and experienced EHPs with long-term associations with the CIEH able to offer public health and legal opinion. These have regularly attracted over 1,000 attendees, and while they are based around the 'question-answer' format, they can 'free-wheel', and, as the name suggests, turn into 'conversations'.

Mention has been made of a listing of EHPs willing to make themselves available to undertake COVID work. On 25 March 2020, via e-mail, members heard directly from the CIEH Director, Northern

Ireland (Gary MacFarlane) that he was co-ordinating a national initiative on behalf of the CIEH in working with stakeholders, including government, to establish a 'register of volunteers'. These would be CIEH members and others '... with environmental health skills, experience and qualifications' who might find themselves '... available to volunteer their services for the public good', by joining the National Environmental Health Volunteer Register (NEHVR). The roles he envisaged included, providing support and advice to businesses, manning local advice lines, and assisting local authorities and port health authorities.

There then followed what looked to be very encouraging signs. Not only did the Register grow (it would eventually have 400 registrants) but in an update on 31 March 2020 (CIEH, 2020d) that the database of volunteers (then 350 strong) would be made available that day '... to a wide range of organisations across the public, private and community/charity sectors' inviting recipients to use it. In addition, members were told that the CIEH was '... in communication with key government departments to suggest that [they] could also assist them'.

Given the enthusiastic support the Register was apparently receiving from partner organisations, including the Public Health agencies in the three devolved nations, there was every indication that suitable registrants might be assigned a role that involved contact tracing. Indeed, believing that this would be a job performed at a local level, and that agencies were now in discussion with local authorities, the directors for Northern Ireland and Wales stated in a further update on 29 April 2020 (CIEH, 2020e) that they felt 'confident that all three will use the skills of EHPs in this area going forward', going so far as to suggest that there were on-going discussions underway with potential partner organisations – the Association of Directors of Public Health (ADPH), the Local Government Association (LGA) and the Society of Local Authority Chief Executives (SOLACE) – '... to design the contact tracing approach across the country'.

However, it soon became clear that this was not what Public Health England, or its parent department, the DHSC, had in mind when they unveiled their centralised system in what would become 'NHS Test and Trace'. As for the NEHVR, though the number of volunteers being taken on to assist local authorities and other organisations was unknown, when asked about this in November 2020, the directors of both Wales and Northern Ireland said that they thought it had been 'disappointing'. This they put down, at least in part, to a reluctance on the part of local authorities to employ volunteers.

If the first effort to compile a register of those available to bring their skills and experience to bear on the pandemic had largely failed, the next followed an announcement by the prime minister in Parliament on

8 September 2020 that it intended to create what the CIEH announced on in a News item the following day (CIEH, 2020f) as '... a register of Environmental Health Practitioners (EHPs) for local authorities in England to draw upon for support in combatting COVID-19'.

What emerged was "Environmental Health Together," a register that would enable '... local authorities in England to access the help they need to combat the pandemic' (CIEH, 2020g). Hosted by the LGA, backed by MHCLG and NHS Test and Trace (though not DHSC, as mentioned previously), it would be professionally quality assured by the Chartered Institute of Environmental Health (CIEH). The register was duly opened at the end of October, and applicants could opt for full- or part-time work, and make themselves available to work in person or remotely, in a remunerated or voluntary capacity. It will be interesting to discover in due course both the level of interest the register generated and its uptake by local authorities in support of their local efforts to protect their communities.

On the use of EHPs for contact tracing, the CIEH has been clear from the moment they launched the NEHVR that it saw a place for EHPs in disease control work, particularly supporting PHE in contact tracing. Realising that the offer of EHPs from its original Register had been passed over, concerted efforts were made by the Institute in May to assist EHPs in their recruitment to the privatised 'NHS Test & Trace' system at either Tier 3 (Contact Tracer) or Tier 2 (Clinical Contact Caseworker), by releasing data from the Environmental Health Registration Board's register and encouraging members eligible to do so to apply. EHPs were duly employed by NHS Professionals to the more senior of the two roles, some, as we now know, working at Tier 1 and with PHE.

Several of the UK EHPs responding to our requests for their reflections on the response to the pandemic passed comment on the contribution that they thought the CIEH had made. They, along with others, were as one in complimenting the CIEH for its 'good work' since April, mentioning the webinars and engagement with the membership and stakeholders via digital communication platforms. However, perceived inaction during the early stages of the epidemic in Britain was a source of discontent, with several members criticising the CIEH for its failure to react early enough and with the strength of purpose to make a difference when COVID-19 arrived. One long-time member (PC8), while acknowledging that hindsight was a wonderful thing, considered the CIEH 'asleep at the wheel' in January and February, when they should have been '... lobbying Government, PHE ... and local authorities'. By so failing to promote the availability of EHOs, the opportunity was lost to deliver '... a public health and environmental health proportionate response'.

As the respondent suggested, it is easy to say so with hindsight (especially so without the background story), but the CIEH was probably later than it should have been in engaging with EHPs over the impending pandemic and championing their cause. This should not be taken as a criticism of the officers and notable practitioner-members who have clearly done a great deal over the months. Their efforts and achievements are there for the record. However, the failure to impress the Government, and Public Health England in particular, early on of the credentials of EHPs to undertake outbreak investigation and contact tracing should be a source of regret to all, not least to those that might have benefited from a bank of trained health professionals supporting PHE and their Health Protection Teams.

6.3 Membership bodies – the Local Government Association

The LGA is a national membership body to which the vast majority of local authorities in England (and Wales, through the Welsh Local Government Association) belong, working, as it says of itself, '... on behalf of our member councils to support, promote and improve local government'. Since the seriousness of the epidemic in Britain became apparent, the LGA has been active in seeking to engage with local authorities on a wide range of COVID-19 matters, and the authors of this book can vouch for their efforts to see Environmental Health and its practitioners recognised as an important element of the national response to the pandemic.

More recently, the LGA has been drawing attention to the importance of considering the entire workforce of local councils as a resource that might be integrated and mobilised locally, arguing that '... not a single area of local government has not been affected by the COVID-19 pandemic'. Its website, and in particular the extensive library of 'information for councils' (LGA, 2021a) which links to 'COVID-19: good council practice', attests to the 'remarkable work' that councils have and are doing '... to address the challenges brought by COVID-19'.

Currently, several entries refer to Environmental Health and/or EHPs, though these are being frequently refreshed and so more are likely to be added in future. In addition, there are some excellent 'sketches' submitted by DsPH describing their work over the last year. These are posted under the title 'Public health on the frontline: responses to COVID-19' (LGA, 2021b).

In addition, the website links to examples of good practice that might be pursued and adopted by other local authorities. However, one, developed by MHLGC, provides a framework on which to build capacity to meet the challenge of COVID-19. This might prove useful to those managing their EH response, giving examples of good practice from councils around the country (LGA, 2021c).

6.4   Academic and Learned Institutions

One historical difficulty encountered by local government EHPs has been to obtain sight of peer-reviewed publications, essential to the performance of evidence-based practice. 'Open Access' has helped immeasurably, but there were often financial and other obstacles, and in a fast-moving situation as that rolling out in the first weeks of the pandemic, this might have impeded informed, timely action. However, the need to share new and reliable information about the virus, disease, and its epidemiology, ideally bought together in one place, was recognised by different organisations, and free access, without restriction, granted to journals through their various publishing platforms.

Countries around the world will have their own preferred sources of academic literature and guidance, so we are reluctant to present those to which EHPs in the UK might choose to consult as especially authoritative.

It can only be hoped that the experience of resorting to peer-reviewed sources through the academic literature, and consulting learned institutions for reliable data on COVID-19, will continue to inspire EHPs to discover the evidence that should inform policy and deliver good practice after the pandemic.

6.5   Ad hoc support for Environmental Health practitioners

Social media has undoubted popularity, and it provided assistance to the authors when looking to obtain a flavour of the 'COVID-19 stories' from EHPs around the world, with LinkedIn providing the names of practitioners for possible future contact.

Two 'groups' that have played a significant part in both allowing for exchanges of views between EHPs on COVID-19, and advocating on their behalf, are described below. These are merely examples – we have no doubt that similar forums exist wherever in the world two or more EHPs are gathered! – though perhaps serving as encouragement to anyone wishing to become involved in these, or as inspiration to others who might be thinking of setting up a group to serve a similar purpose.

6.5.1 Sharing good practice: the UK COVID-19 Knowledge Hub

Since the start of the pandemic the ability to share environmental health knowledge and experience has become a critical part of our national COVID-19 response work. For UK local government EHPs, this has been assisted since March 2020 by the creation of the local authority environmental health, licensing and trading standards COVID-19 collaboration on the Knowledge Hub platform (https://khub.net/).

Coordinated by EHPs from the North East Public Protection Partnership (NEPPP), it has become one of our first 'ports of call' whenever questions arise, particularly around interpretations of UK legislation and guidance and potential regulatory approaches. The KHUB is also used by participants to spread the word about interventions, particularly where local events potentially have national implications.

At the time of writing the Hub had nearly 2,800 members and recently extended its reach to include police officers who, on account of their increasing involvement in COVID-19-related law enforcement work across the UK, could potentially benefit from access to the broad source of public health knowledge that EHPs and others can provide.

6.5.2 Advocating for EHPs: the UK Chief Environmental Health Officers Group

In 2013, the Chartered Institute of Environmental Health (CIEH) created a National Environmental Health Board (NEHB) to build relationships between local government environmental health professions, other regulatory and public health organisations, and national government departments. The then CIEH chief executive, Graham Jukes, wanted the Board to ensure '… environmental health capacity is factored into government department and agency thinking' and to facilitate '… a more joined-up and co-ordinated response to issues such as food fraud, flooding or significant outbreaks of infectious disease'.

The NEHB ceased functioning a few years later, but former members of the NEPPP decided to create a new Chief Environmental Health Officers group recruiting others from more than 100 local authorities across England, Scotland, Wales, and Northern Ireland. They first met in London in January 2019 as the 'Future of Environmental Health Group' and began working on areas of national concern, including the environmental health workforce and forthcoming changes to environmental health qualifications. After a second meeting in early 2020 their name changed to the UK Chief Environmental Health Officers Group, though, needless to say, COVID-19 has changed their focus.

In a short time, the Group has become the de facto voice for local government EHPs on national government COVID-19 policy. Its members and others, including CIEH policy officers and Environmental Health academics, continue to inform government department and agency thinking. This is an extraordinarily complex and political process, with EHPs only one voice among many. However, the UKCEHO Group, led by Peter Wright, has helped to make more visible the role of EHPs in the eyes of national government departments, arm's length organisations and the wider Public Health profession. Furthermore, its role illustrates the continuing need for locally informed Environmental Health thinking in national pandemic policy processes. At a time when EHPs in the UK found themselves having to respond individually to the unfolding crisis, Peter and the Group have seen to it that their experiences from 'the coal face' have been brought together and presented as the 'voice' of Environmental Health.

References

Chartered Institute of Environmental Health (2020a). Press release: *CIEH issues coronavirus guidance* by Ross Matthewman (6 March 2020). Available at: https://www.cieh.org/news/press-releases/2020/cieh-issues-coronavirus-guidance/ (accessed: 23rd February 2021)

Chartered Institute of Environmental Health (2020b). EHN article: *CIEH: Supporting you in the fight against coronavirus* by The CIEH Team (19 March 2020). Available at: https://www.cieh.org/ehn/public-health-and-protection/2020/march/cieh-supporting-you-in-the-fight-against-coronavirus/ (accessed: 23 February 2021)

Chartered Institute of Environmental Health (2020c). EHN Article: *Use us to fight the virus. It's our job* by Katie Coyne (19th March 2020). Available at: https://www.cieh.org/news/press-releases/2020/cieh-issues-coronavirus-guidance/ (accessed: 23 February 2021)

Chartered Institute of Environmental Health (2020d). News items: *Update on CIEH's COVID-19 National Environmental Health Volunteer Register* by Gary MacFarlane (31st March 2020). Available at: https://www.cieh.org/news/blog/2020/update-on-ciehs-covid-19-national-environmental-health-volunteer-register/ (accessed: 23 February 2021)

Chartered Institute of Environmental Health (2020e). News item: *Update on national volunteers register, contact with public health agencies and contact tracing* by Gary MacFarlane and Kate Thompson (29 April 2020). Available at: https://www.cieh.org/news/blog/2020/update-on-national-volunteers-register-contact-with-public-health-agencies-and-contact-tracing/ (accessed: 23 February 2021)

Chartered Institute of Environmental Health (2020f). News item: *CIEH welcomes Prime Minister's support for environmental health in combatting COVID-19* by Ross Matthewman (9th September 2020). Available at: https://www.cieh.org/news/press-releases/2020/cieh-welcomes-prime-ministers-support-for-environmental-health-in-combatting-covid-19/ (accessed: 23 February 2021)

Chartered Institute of Environmental Health (2020g). News item: *Environmental health together launched to combat COVID-19* by Ross Matthewman (9th November 2020). Available at: https://www.cieh.org/news/press-releases/2020/environmental-health-together-launched-to-combat-covid-19/ (accessed: 23 February 2021)

Dyjack, D.T. (2020). *Open letter to the environmental health profession* from the CEO of the National Environmental Health Association (9 April 2020). Available at: https://www.neha.org/news-events/latest-news/open-letter-environmental-health-profession (accessed: 23 February 2021)

Environmental Health Australia (2020a) – Environmental Health Officers and their role in the community in response to COVID-19 (9 April 2020). Available at: https://www.eh.org.au/documents/item/1131 (accessed: 27 April 2021)

Environmental Heath Australia (2020b). *Environmental health professionals around Australia – COVID-19 pandemic report*, made available via IFEH (https://ifeh.org/) at: https://ifeh.org/covid19/docs/Environmental%20Health%20Professionals%20Australia%20-%20COVID-19.pdf (accessed: 23 February 2021)

International Federation of Environmental Health (2020a). IFEH Update: *Coronavirus – COVID-19* (8 March 2020). Available at: https://www.ifeh.org/docs/ifeh_messages/COVID-19_IFEH_Update.pdf (accessed: 23 February 2021)

International Federation of Environmental Health (2020b). *Briefing from IFEH President* (Professor Susana Paixão) in December 2020. Available at: https://www.ifeh.org/docs/ifeh_newsletter/IFEH_Newsletter_Dec_2020.pdf?fbclid=IwAR1lwQ9OJhGjdDYWjv3-aF1GF9HOkxgEMscknduYSMjC93fRQ1BCd-eAFX8 (accessed: 30 December 2020)

Local Government Association (2021a). *Coronavirus: Information for councils* (website) at: https://www.local.gov.uk/our-support/coronavirus-information-councils links to: *COVID-19: good council practice*. Available at: https://www.local.gov.uk/covid-19-good-council-practice (accessed: 12 February 2021)

Local Government Association (2021b). *Public health on the frontline: Responding to COVID-19*. Available at: https://local.gov.uk/our-support/coronavirus-information-councils/covid-19-service-information/covid-19-public-health-0 (accessed: 24 February 2021)

Local Government Association (2021c). *Local authority COVID-19 compliance and enforcement good practice framework, January 2021*, produced by the Ministry of Housing, Local Government and Communities. Available at: https://www.local.gov.uk/local-authority-covid-19-compliance-and-enforcement-good-practice-framework-january-2021 (accessed: 23 February 2021)

Mudalay, S. (2019). *International Federation of Environmental Health Past President's Report – 2018–2020*. Available at: https://www.ifeh.org/docs/news/PAST%20PRESIDENT.pdf (accessed: 23 February 2021)

National Environmental Health Association (2020a). *COVID-19 Environmental Health Workforce Rapid Needs Assessment Report.* Available at: https://emergency-neha.org/covid19/wp-content/uploads/2020/11/NEHA_COVID-19_EH_Workforce_Rapid_Assessment_Report.pdf (accessed: 5 January 2021)

National Environmental Health Association (2020b). *COVID-19 Environmental Health Workforce Needs Assessment Report II.* Available at: https://emergency-neha.org/covid19/wp-content/uploads/2020/11/COVID-19-EH-Workforce-Needs-Assessment-II-Report.pdf (accessed: 5 January 2020)

Swain, P (2020) - An open letter to Ministers, Health and other Government Departments throughout Australia from the National President of Environmental Health Australia (22 March 2020). Available at: https://www.eh.org.au/news/covid-19-an-open-letter-to-ministers (accessed: 27 April 2021)

7 Reflections on the global Environmental Health response ... so far

7.1 Introduction

This chapter contains points that that emerged during the compiling of the book that, though they would not have fitted into any of the previous chapters, seemed too important to omit.

No doubt, if we were to come back to this in six months, and asked what practitioners were doing and what they had achieved in the first nine months of 2021, we would have a quite different story to tell. However, we consider that the, sometimes, chaotic response to challenges faced in the early months of the first wave of the pandemic, even now, in January 2021, with many issues outstanding or not yet properly resolved, has produced a valuable snapshot from which to learn lessons right now.

7.2 The big picture

The consensus view is that Environmental Health practitioners have made a very considerable contribution world-wide to the public health response to COVID-19.

The then president of the International Federation of Environmental Health, Dr Selva Mudaly, announcing the theme of World Environmental Health Day on 26 September 2020 as, 'Environmental Health, a key public health intervention in disease pandemic prevention', cited and quoted from Morse et al. (2020) on the contribution made by EHPs in the control of COVID-19 in sub-Saharan Africa. Here they refer to the 'vital role' that practitioners had performed since the outset, many of which were mentioned in earlier chapters.

Though by no means complacent – they acknowledge'... health systems crippled by the delivery of routine health services and where COVID-19 mitigation measures such as social distancing and lockdowns are neither physical or economically viable' – the Morse group were conscious that Africa was well prepared to deal with a pandemic because it

DOI: 10.1201/9781003157229-7

had '… learnt and developed plans from previous and ongoing disease outbreaks', citing, here, Ebola in 2014–2016.

However, they believe it could be argued that given the breadth of core functions and depth in specialisms focussed on preventive health, and especially their '…overarching, holistic range of skills' EHPs had '… a much wider role to play' in this and future pandemics, in sub-Saharan Africa.

The suggestion that EHPs had undertaken and accomplished a great deal during the opening foray with the infection, but could do more, is a conclusion drawn by others and one that we will return to in the final chapter.

7.3 The Environmental Health response

The COVID-19 pandemic has provided EHPs with an opportunity to demonstrate their skills and qualities like no other situation before. As an occupational group they are not unique in this, though it is hard to imagine many other health professionals finding themselves being engaged in so many diverse activities. This has undoubtedly revealed them to be a significant asset to the public heath workforce at time of emergency or crisis, that is, when they have had an opportunity to demonstrate their capabilities.

What is clear from our respondent group is that, with some notable exceptions, they felt that the COVID-related duties and activities they were allocated were well within their capabilities, but they could have done more and sooner. Among the limiting factors were their national government's slow realisation that in Environmental Health they had a viable resource with the skills and expertise to support, if not lead, the Public Health response.

Excluding those in a position to offer their services voluntarily, or in a consultancy capacity in the private sector, the extent to which practitioners might have the freedom to act at will or impulse, or to show initiative, was clearly dependent on seniority and/or willingness on the part of employers to see their staff deployed to work in a different capacity. Otherwise EHPs are in public service and perform tasks at the behest of their employers, who, in turn, may be bound by their governments to fulfil a range of duties and responsibilities. Under normal circumstances their attention would be focussed on the hazards and risks presented in the normal course of their work, but, here, this shifted at pace to a new hazard presenting a risk in multiple settings, in this case infection with SARS-CoV-2.

Before considering where shortcomings cropped up, and how these might be remedied for the future, we should acknowledge those public sector employers who, recognising early on that they had an asset in their Environmental Health workforce, saw to it that practitioners were put to good use in a capacity suited to their knowledge and skill base. Where this was the case, managers acted swiftly and decisively, and special operational plans were adapted or drawn up that had the effect of altering or overriding standard operating procedures.

7.4 Preparedness to respond and deploy

Countries around the world would have identified a critical moment or series of moments when their government took heed of the situation and initiated decisive public health action. In some cases, this action was immediate and extreme, involving the closing of all land, sea and air borders, and the forced quarantining of cases and contacts. However, with several notable exceptions, countries boasting more liberal administrations were reluctant to impose Draconian measures, instead opting for targeted guidance on travel from particular countries, and appealing to the public to adopt hygienic behaviours, including social distancing, regular hand washing and mask wearing.

Some of those responding to our invitation to share their 'story' spoke of being involved early on in putting into action the most fundamental public health measures designed to limit the spread of infectious disease, indicating their country had been quick to see the nature and extent of the problem and the value of mobilising the wider public health workforce. For the remainder, countries found themselves moving inextricably towards 'lockdown' and the suspension all international travel, restrictions on movement other than for essential purposes, and closing down of all but essential commercial operations, and the freedom to congregate with others. Then, following a period dictated by politicians, but supposedly advised on by epidemiological guidance, liberties and livelihoods were permitted to resume.

Environmental Health, and its front-line practitioners, managed to play a part in all three of these phases of the 'first wave' of the pandemic, closing borders and assisting some business operations to shut down and safely remain 'moth-balled', assisting others to stay open but safely, and then overseeing the 'unlocking' of the rest, again, looking to safeguard both staff and customers in the process.

7.5 Maintain 'business as usual' or 'seize the day'?

Although our interest was naturally focussed on activities performed by EHPs in limiting the spread of infection, we should not overlook the fact that some practitioners continued, in whole or part, to undertake their regular duties and responsibilities, either by adding hours to their working weeks or by finding 'smarter' ways of doing their routine work, thus allowing them the time to take on COVID-related activities.

Our survey has indicated that some public sector employers might have been slow to take EHPs off their routine work and redeploy them onto COVID duties, but once they did, it become their predominant work activity. We cannot tell at this stage whether every local authority got the balance right, but there was sufficient indication in the responses received that EH managers recognised that by abandoning their normal duties and responsibilities harm might come to those affected.

Whatever might have been the hope that environmental health departments and their staff would be able to continue 'business as usual', as it became obvious that SARS-Cov-2 wasn't going to conform to type, and, like SARS and MERS, fizzle out, there was a demand to bring all necessary and available resource to bear on the disease in all affected countries around the world. As we have seen, in some countries, Environmental Health was seen as an available source of expertise and manpower, and EHPs were slotted into – if they were not there already – the wider Public Health workforce. In others, this was not the case until it became clear that there was no one else available or prepared to go into the community to see control legislation implemented and guidance upheld.

Some EHPs, particularly in the early weeks of the emergency, seem to have quickly shrugged off their natural reticence to put themselves forward and chose to 'seize the day', becoming self-aware of strengths and skills, perhaps for the first time in their careers. Then, by taking on duties and activities that must have challenged them technically and professionally, and so taking them out of their 'comfort zone', demonstrated to others the nature and extent of their worth.

As has been commented on before, this came with some difficult decisions for managers to take, not least how to avoid greater harm arising from the discontinuation of their regular work when considered against their new 'COVID work'. If this presented a 'difficult call' when the numbers of cases, hospital admissions and COVID-19 deaths were low, it seems to have been much less so when there was urgent need for local government staff to enforce restrictions on business activity ahead of 'lockdown', be they legislative measures or delivered through guidance.

Here we see one of those moments in the pandemic where EHPs have been seen, sometimes late in the day, as the 'natural choice' for delivering a wide range of infection control measures, and that on reflection, this delay may have come late, and, possibly, at immeasurable cost.

7.6 How COVID-19 exposed the diminished Environmental Health resource

Though it is impossible to say whether this is common to all countries of the world that turned to its Environmental Health workforce to respond to COVID-19, its reduced capacity over time was mentioned often by respondents to the survey, indicating that, when 'the chips were down', there were insufficient personnel available to meet the call, thus a limiting factor in terms of the speed and extent of the Environmental Health response.

As far as preparedness is concerned, and especially the ability to act quickly and decisively when challenged by a novel coronavirus, COVID-19 has exposed significant gaps in public health infrastructure, most starkly in those countries that, 20 years ago, might have been able to hold their systems up as a model. Here in the UK, though as we have heard elsewhere but not everywhere, these gaps are the result of a sustained period of cuts in public sector spending following the banking crises in 2008/09, and where no case has been made, or could be found, to protect the funding available to Environmental Health (Dhesi, 2019).

7.7 How the COVID-19 pandemic shone a light on health inequality and inequity

It is widely accepted that the pandemic has highlighted, indeed magnified, inequalities (and inequity) in health especially in respect of the BAME communities that have suffered disproportionately. In response to what has become a 'mantra' for Britain as it emerged from the first wave of the pandemic of 'build back better', Michael Marmot's plea to 'Build Back Fairer' (Marmot et al., 2020) argues that the levels of social, environmental and economic inequality in society were damaging enough to health and well-being prior to the pandemic striking, citing his team's previous work described in Chapter 4.. However, as 'Build Back Fairer' maintains, '… inequalities in mortality from COVID-19 and rising health inequalities as a result of social and economic impacts, have made such action even more important'.

Worse still, the pandemic has further exacerbated the inequity identified by Marmot where the impact of COVID-19 has been felt most acutely by those on low incomes and living in poor conditions especially those vulnerable through age, and as has become all too apparent, those living in residential care homes. Although this might have been expected, the pandemic has had a far greater impact on the BAME groups than the remainder of the population here in Britain, and, we suspect, elsewhere. Not only have they faced and succumbed to the infection in healthcare settings, both as patients and employees (Farah & Saddler, 2020), but they have been the most affected group of those in disadvantaged positions. As a consequence, it has been impossible not to see how health and societal inequalities are mirrored in those lost to COVID-19 (Johnson, Joyce & Platt, 2021).

While the worsening in deprivation in UK is attributed, in part, to financial austerity in the last decade, the key role performed by EHPs in tackling health inequalities, not least in ensuring safe and healthy living and working conditions, and protecting vulnerable people in society, must be recognised and resourced (Dhesi, 2013). The failure of some local authority environmental health departments to maintain the necessary housing enforcement resource – one recent study found that some local authorities had no environmental health officers dealing with housing – is evidence of this (Battersby, 2018). It is vital that this longer-term role is not overlooked in responding to the immediate pressures of the COVID-19 pandemic.

7.8 Environmental Health comes of age

Had EHPs found themselves only engaged in business compliance work this book would have been simple to compile but a pale celebration of their contribution to public health. At its heart has been the realisation – rapid in some countries but painfully slow in others – that they are 'in touch' with how people in their local populations live, work, and socialise.

Of course, under normal circumstances, what causes EHPs to interact with real people in the real world is, in the main, humdrum and unremarkable; they visit and inspect their houses, workplaces, local shops and pubs. However, in most they encounter and engage with worried tenants, misinformed employers, and confused shopkeepers, and in so doing develop a true understanding how people 'tick'.

While one cannot disregard the importance of empathy, and desire to act on behalf of others less fortunate and who might not have a 'voice', these are difficult attributes to assess. Yet behind this image of a local authority official performing largely routine, reactive work is the health professional expected to respond to serious and unexpected situations that require the ability to think clearly and act decisively. In this mode, they might be expected to organise and lead investigations, develop and manage interventions and respond to emergencies.

As a profession, those bodies responsible for registering and endorsing practitioners to practise EH have held to the broad notion that first and foremost EHPs should be possessed of a technical competence derived from principles underpinned by science. Although there will be variations on matters of detail, those regulating Environmental Health qualifications recognise the importance of 'preparation for practice' which is a product of higher education, preparedness to undergo practical and professional training, and devotion to the principle of continuous professional development.

It is now, here on record, that EHPs throughout the world have brought together their knowledge and understanding of infectious disease and its transmission; used this in the most fundamental public health tool of 'contact tracing'; demonstrated their ability to translate often complex issues around the virus and COVID-19 into clear and concise messages; and shown their special dedication in taking these messages, and thus public health, to the people they serve.

If this wasn't evidence enough of a profession finding its maturity, EHPs have often had to deal with some challenging situations where they encountered dissent and sometimes outright opposition. Putting aside the risk to their own health from the virus, one is struck by the willingness to repeatedly put themselves through this when the behaviour they are attempting to change may be underscored by deep-rooted cultural or spiritual beliefs, confused, still further, by 'COVID-19 myths'.

7.9 Environmental Health answers the call ... but was it the right call?

When asked about their COVID work, many EHPs described being preoccupied with legislation and guidance to restrict commercial and recreational activity that might cause onward transmission. The sheer volume of this, its frequent amendment and re-issue, and the need to distinguish guidance from statutory requirement were often the greatest challenges for EHPs. This might be explained by the fact that routine EH practice is popularly characterised as EHPs performing a regulatory enforcement role at the expense of any other, and as we have discovered in researching this book, overlooking the better use of their wider skills in tackling the many and various challenges of the COVID-19 pandemic.

That so far there hasn't been widespread dissent about this is to the credit of EHPs; but is this an example of a presumption made throughout the pandemic that they should only be assigned to compliance-related duties? To continue to use EHPs as 'sanitary policemen' is arguably to continue to overlook and underuse the depth and breadth of their public health abilities.

7.10 A clear 'force for good' ... but could EHPs be doing more?

As we were bringing this book together , a paper was produced by a multi-national group of Environmental Health academics (Rodrigues et al., 2020) that provided a means of cross-checking our data. By any estimate they are in alignment, almost to the extent that the paper's Abstract could have been written for this book. While this should not be surprising given their similar remits, we detail below a slightly abridged version of their Abstract:

> The COVID-19 pandemic highlighted the relevance of public health professionals all over the world, in particular EHPs, who played a major role in the containment of the novel coronavirus, SARS-CoV-2. However, as in past disasters, their involvement was orientated towards urgent tasks, and did not fully utilise EHPs' competencies and skills. Additionally, due to limited resources, during emergencies, EHPs may temporarily transition away from their day-to-day role, potentially increasing other public health and safety risk factors without appropriate surveillance or intervention.
>
> Findings from discussions concluded that, despite the observed differences across the countries, EHPs are one of the main public health emergency preparedness and response actors. However, since resources are still lagging significantly behind need, we argue that the role of these professionals during pandemics should be focused on practices that have higher impact to support population health and safety.

It is not intended to dissect the paper here since readers can readily and freely obtain a copy as it is available 'open access'. For now, we echo the sentiments of Matilde Rodrigues's group in their conclusions where, having acknowledged the differences in the interventions performed by EHPs in different countries during the pandemic, they believe that '... lessons from the experiences in different countries provide relevant information about the critical roles of EHPs in responding to, monitoring and controlling the risk in population safety and health during a pandemic and in its aftermath'.

References

Battersby, S.A. (2018). *Private rented sector inspections and local housing authority staffing* (supplementary report prepared for Karen Buck MP). Available at: http://www.sabattersby.co.uk/documents/Final_Staffing_Report_Master.pdf (accessed: 26 May 2021)

Dhesi, S. (2013). Reflexivity – Researching practice from within. *Journal of Environmental Health Research*, 13(1), 83–87.

Dhesi, S. (2019). *Tackling Health Inequalities: Reinventing the Role of Environmental Health*. Oxon: Routledge.

Farah, W., & Saddler, J. (2020). *Perspectives from the front line. The disproportionate impact of COVID-19 on BME communities*. NHS Confederation, BME Leadership Network. Available at: https://www.nhsconfed.org/-/media/Confederation/Files/Publications/Documents/Perspectives-from-the-front-line_FNL_Dec2020.pdf (accessed: 24 February 2021)

International Federation of Environmental Health (2020). Announcement by Dr Selva Mudaly of theme of *World Environmental Health Day* – 26 September 2020. Available at: https://ifeh/wehd/2020/WEHD2020.pdf (last accessed: 30 December 2020)

Johnson, P., Joyce, R., & Platt, L. (2021). *The IFS Deaton review of inequalities: A new year's message* (5th January 2021). Available at: https://www.ifs.org.uk/inequality/the-ifs-deaton-review-of-inequalities-a-new-years-message/ (accessed: 24 February 2021)

Marmot, M., et al. (2020). *Build Back Fairer: the COVID-19 Marmot Review*. Institute of Health Equity. Available at: http://www.instituteofhealthequity.org/resources-reports/build-back-fairer-the-covid-19-marmot-review/build-back-fairer-the-covid-19-marmot-review-full-report.pdf (accessed: 23 February 2021)

Morse, T., Chidziwisano, K., Musoke, D., Beattie, T.K., & Mudaly, S. (2020). Environmental health practitioners: A key cadre in the control of COVID-19 in sub-Saharan Africa. *BMJ Global Health*, 5, e00314. https://doi.org/10.1136/bmjgh-2020-003314. Available at: https://gh.bmj.com/content/bmjgh/5/7/e003314.full.pdf (accessed: 11 February 2021)

Rodrigues, M.A., Silva, M.V., Errett, N.A., Davis, G., Lynch, Z., Dhesi, S., Hannelly, T., Mitchell, G., Dyjack, D., & Ross, K.E. (2020). How can environmental health practitioners contribute to ensure population safety and health during the COVID-19 pandemic? *Safety Science*, 105136 (https://doi.org/10.1016/j.ssci.2020.105136R). Available at: https://www.sciencedirect.com/science/article/pii/S0925753520305336 (accessed: 14 February 2021)

8 Learning lessons from the global Environmental Health response to COVID-19 so far – conclusions and recommendations

8.1 Introduction

We set out to discover and report on the nature and extent of the contribution made by Environmental Health Practitioners in meeting the challenge of COVID-19, and how this had impacted on their normal, day-to-day, work in the early months of the pandemic, hoping, along the way, to secure an insight into their thoughts, fears, failures and achievements. We offer the following conclusions and recommendations to enhance the contribution of Environmental Health during what remains of this pandemic, and to prepare better for next time. We also look beyond the pandemic and consider how Environmental Health should build for a future made clearer by COVID-19, when EHPs can return to improving, as well as protecting, health.

8.2 Regaining a 'visible' Public Health presence for Environmental Health

EHPs often felt passed over for undertaking more responsible COVID-19 tasks because their professional background and experience had gone unnoticed and unappreciated by leaders locally, regionally, and nationally. EHPs spoke of others, bearing titles that at the time were unfamiliar, performing – in some cases, under-performing – activities more suited to their own skills.

Whether a departmental name or job title needs to bear witness to Environmental Health is a moot point, but it might to a senior manager unfamiliar with the disciplines under her management, or the director of another service delivered via a different tier of local government. Over time, Environmental Health's health role may have become so unfamiliar to those considering long-term regional/national strategies that it is overlooked.

In this respect, and only this respect, the COVID-19 pandemic has proved beneficial to Environmental Health and EHPs as practitioners

DOI: 10.1201/9781003157229-8

on the ground. Through their efforts, effective Environmental Health controls have been put in place, thus reducing the spread of disease, and saving lives. They report how the professional discipline, and their own contributions, have been recognised locally by politicians and other health professionals who may not have appreciated their worth before. This, then, is the moment when Environmental Health can and should maintain the focus on its practice.

Recommendation: Given that the pandemic is set to continue into 2021 and beyond, EHPs should take every opportunity to maximise their contribution to the effort by making full use of their talents. Therefore, alongside their daily COVID-19 work, they should explore where opportunities might exist for them to take on greater responsibility, lead projects and develop initiatives. Against this is the need for those already heavily committed not to 'burn out' and look out for warning signs that they may be 'overdoing it'. Staying 'mentally safe' might be just as important as staying so physically.

8.3 Establishing the best point of delivery for public health measures

The pandemic has drawn attention to the performance of bodies required to oversee public protection at governmental level throughout the world. In England, for instance, Public Health England has been shown to be under-resourced and exposed when there was a need for both urgent assessment of high-level risks and priorities, and the need to undertake front-line, public health work, including outbreak investigation and 'contact tracing'.

There remains a case for maintaining close integration of these different elements that make up the national infectious disease response, but COVID-19 has demonstrated they do not have to be performed by the same agency. Indeed, all the indications are that roles such as outbreak investigation and contact tracing can be performed very effectively and efficiently by local teams.

Recommendation: Governments are urged to review responsibilities for the planning and delivery of the national response, taking into account the vital response roles undertaken by the local government Public Health workforce.

8.4 Seeing and being seen – research as the 'torch to visibility'

One message from this book is that, as individuals, EHPs have achieved more than they might have previously given themselves credit for, while other health professionals have seen, perhaps for the first time, a body of skills honed by experience and a capability to deliver interventions that might otherwise have failed to achieve their purpose. While 'doing' is normally far more important than 'talking about doing', here is an opportunity to do both. We have no doubt that EHPs will continue to use their talents more and more if they choose (or are permitted) to do so. However, in addition, and setting aside their natural reticence to 'blow their own trumpet', the same EHPs should look to do what the authors of this book have sought to do in representing the work performed by EHP during the pandemic so far, by recording what they are doing, reflecting on it, and evaluating its impact.

Recommendation: All EHPs are urged to become researcher-practitioners, to write about their experiences, to reflect on what they and their colleagues have done, to consider whether it has worked (or not) and what implications it has for the future, then – and this is the most important aspect of the recommendation – see that it is read by those in their social media network or magazine circulation, and, if at all possible, see it submitted to a journal for publication (be it a paper, letter or commentary) so that it can be read and cited by other health professionals. The 'power of the pen' should never be overlooked, and the guidance set out in one of the UK Environmental Health Research Network's earlier publications 'Evidence, research and publication: a guide for EHPs' (Couch et al., 2012) is free to download, and remains as relevant as ever. At the same time, we call on employers to create an environment to support these endeavours, since they will ultimately contribute to the evidence base of interventions effective in protecting the public's health.

8.5 How COVID-19 caused EHPs to 'win friends and influence people'

The multi-disciplinary and multi-agency responses to COVID-19 have brought together health professionals that may never have worked together before, particularly those more focussed on health promotion, or else primarily involved in surveillance and epidemiology. Thus, EHPs were able to draw from their experience and expertise, while demonstrating their own skills and abilities, thus asserting the case for using risk-based and holistic approaches to complex problems.

Recommendation: Every effort should be to maintain and build on the relationships created between practitioners and agencies after the pandemic, to their mutual advantage and benefit to public health.

8.6 Building on the global Environmental Health experience

The International Federation of Environmental Health has done a great deal over the past nine months to put its refreshed vision and mission into practice by making available material submitted by its members on their respective COVID-19 responses. We applaud those responsible in its governance for swiftly recognising the role that they might play in mobilising EHPs around the world, but especially in the countries represented by its membership. However, if the full value of global representation of Environmental Health as a professional discipline is to be realised, the IFEH needs to be able to speak for as large a constituency as possible.

Recommendation: The International Federation of Environmental Health has set itself a challenging series of objectives, reflected in the words of its president, Professor Susana Paixão, in the December 2020 newsletter: '… continue to give worldwide visibility to Environmental Health and attract more members, publicising their work and involving IFEH in more international working groups with key global partners, in particular, with official bodies … essential for the global recognition of Environmental Health workforce' (IFEH, 2020). However, it is beholden on the professional bodies currently members of the IFEH, and by association, every EHP so represented, to become ambassadors for the Federation in building its membership, and seeing it thrive. Environmental Health may be a 'small cog in a big wheel' nationally, but its reach and influence globally can only be enhanced by a Federation representing every practitioner of this discipline.

8.7 Reaffirming the credentials of Environmental Health after COVID-19

All attention for now is focussed on COVID-19 and is likely to remain so for the foreseeable future. However, as practitioners and observers of Environmental Health have pointed out throughout the pandemic, poor housing, unsafe workplaces and unsatisfactory food premises have not gone away because of the infection. It is to the great credit of all involved in delivering Environmental Health services that they have managed to strike some kind of balance between responding to the fluctuating challenge of the virus and their duties protecting the public from other harms, even if this has meant only responding to complaints and dealing with emergencies.

Perhaps at the outset of the pandemic one of the hardest tasks for Environmental Health managers was to establish a post-COVID-19 way of working, given the disruption to the service since March 2020. Yet some very positive signs emerged from our research that would suggest this might not be as problematic as first imagined, given the benefits shown of remote working, greater practitioner control over workload, freedom to work outside 'normal office hours' and so on.

However, this moment is already attracting the attention of central government to review and reorganise the local government platform. For example, in Britain, this is a topic of current interest and concern, particularly given the long-standing tension and distrust between the two. Howsoever this plays out on the wider national and international stages, it is vital that Environmental Health is not compromised by a drive to 'slim down' the public sector, particularly when the pandemic has illustrated so powerfully the vulnerabilities of those already experiencing the worst environmental health. EHPs, given the resources, have an important role in addressing this inequity.

Recommendation: The Environmental Health community, but especially those representing practitioners and whose remit is public health, must do their utmost, now, to impress on politicians and administrators the need to facilitate an ordered return to normal working as the pandemic will allow, but to do this with the clear indication that the priority must be to protect those members of the BAME communities at greatest risk. Thereafter, the focus must return to the longer-term, and more intractable challenge still, of improving health for all and reducing health inequalities, but with renewed vigour. We would encourage governments to follow the lead provided by the World Health Organization in 'Health 2020' through the implementation of 'whole-of-government' and 'whole-of-society' approaches, which secures policy that addresses '… the social gradient in health directly, where interventions are proportionate to the level of health and social need', and focusses on those most affected (WHO, 2020).

8.8 Maintaining the rigour of the qualification

Those professional bodies and others with custody of the qualifications to be held by EHPs wishing to practise or seek registration will likely be looking at their curricula and requirements for practical and professional training in light of the pandemic and considering whether changes need to be made or features reinforced.

In the UK, the Chartered Institute of Environmental Health was at a pivotal point in 2020 in making fundamental changes to the route to professional qualification. Proposed changes included the removal of the need to complete an accredited degree programme, written professional examinations and a portfolio of professional practice and the dissolution of the Environmental Health Registration Board (EHRB). In its place a professional standards pathway for graduates was developed that could be completed within two years and was considered by CIEH to be more in accordance with the qualities and competencies demanded by employers, and in response to wider concerns about the challenges and costs of qualification of the old route. At the time of writing these changes remain in flux, but the pandemic has seen employers demanding proof of EHRB registration as a pre-requisite qualification for COVID-19 work.

The case for ensuring that graduates in Environmental Health emerge with a sound technical education remains clear, but probably supplemented with appropriate skills-based training on matters highlighted by the pandemic. It was especially interesting to receive a call from a UK academic for courses to be 'internationalised', arguing that the focus of domestic courses was too inward-looking, and that international students brought energy and a '... refreshing perspective to Environmental Health issues'.

Recommendation: In light of the COVID-19 pandemic, bodies overseeing the training and qualification of EHPs for practice should look to continue to require candidates to acquire a sound theoretical and practical grounding in Public Health, demonstrating competence in the investigation of an outbreak of infectious disease and other incidents presenting an immediate risk to health. This should be detailed in a curriculum that is reviewed and placed in the public domain. Finally, the registration of practitioners who have satisfied their peers that they are competent to practise should be considered a vital element of this profession as it is to other health professions.

8.9 Enhancing the EHP skill base in epidemiology

It seems likely we will be living with COVID-19 for years to come, as well as its many consequences. Consequently, there will be continued demand for health professionals with the multi-disciplinary skills of EHPs to implement the control measures described above. This is especially important given the likelihood of future pandemics, and the urgency of responses to other imminent global challenges, notably the climate emergency.

Though EHPs were generally satisfied that their professional qualifications fitted them for the challenges presented by COVID-19, criticism was levelled at some courses for the adequacy of their coverage of epidemiology and infectious disease control. While a detailed comparison of curricula and syllabi used at colleges and universities worldwide might be a suitable subject for a research project, we can only invite professional bodies to consider this when they come to accredit institutions delivering qualifying programmes.

Recommendation: EHPs that had chosen to supplement their understanding of the theory and practice of epidemiology spoke of feeling better prepared to deal with the public health challenges presented by COVID-19. If this is the case, and revisions to the qualifying course are impracticable, professional bodies might wish encourage practitioners to study for a post-graduate qualification, such as the Master of Public Health degree, and to offer short courses to members as part of their continuous professional development programmes.

8.10 Governments must equip EHPs with the essentials

In Chapter 3, Morse et al. (2020) argued that the unique role of EHPs in the prevention and control of diseases such as COVID-19 across sub-Saharan Africa justified why governments should recognise, support and invest in the work of EHPs across the region. We think their argument that EHPs should become a 'lynch pin by which preventive and control measures are implemented' and one that leads to a risk-based, trans-disciplinary and local approach to provide 'context appropriate intervention development' is a persuasive one that applies globally. Future interventions will require the development of methods that can reach large numbers with limited resources and at pace. We would endorse these observations and repeat in our Recommendation the Morse team's suggestions as to how this might be delivered, since, as they say, 'the success of this approach requires recognition, investment and support of the EHP workforce', which, readers will recall, accords with the 'essentials' that respondents of our survey said would have allowed them to do more, and to do it better.

Recommendation: That governments make suitable provision such that the EHP workforce going forward is '... effectively trained, supported and provided with the necessary materials and tools such as transportation, communication, protective equipment and educational materials to engage with their communities'.

8.11 Increasing the permanent bank of Public Health professionals

The frequent observation that more could have been achieved with more people on the ground, might be solved in the short-term by the recruitment of fit and able people, suitably trained, to do the policing of COVID-19 controls and restrictions; one thinks here of 'COVID-Marshals' and the redeployment of local government staff who might be precluded from their regular work due to the pandemic.

Beyond this, countries of the world will no doubt be giving careful thought as to how their health protection systems should be configured in the long term, especially looking to improve surveillance and having plans in place to action and co-ordinate an emergency response to newly emergent infections. One would expect this to lead to more graduate training of specialist practitioners capable of performing the tasks thrown up by the COVID-19 pandemic. In the UK this might require no additional support from Environmental Health locally, as local Directors of Public Health can be expected to use their professional body to articulate this when the time comes.

Although the COVID-19 pandemic has brought about unprecedented interest in the aspect of Public Health to do with health protection, it performs several other key roles in seeking to promote and improve health. Here, the focus is on addressing those longer-term factors that impact both positively and negatively on health, and which play out in communities and the population at large.

Efforts to address the social determinants of health, while set aside for the duration of the COVID emergency, have the added imperative that the pandemic has caused inequalities to widen further, as well as inflicting wider harms including the delayed diagnosis of potentially fatal diseases, and support for those with mental health complaints.

Recommendation: Governments are encouraged to unleash the full potential of preventative approaches by committing to training and employing more non-medical health graduates and considering the establishment of a dedicated professional pathway for EHPs to migrate into wider Public Health careers.

8.12 Re-evaluating the investment in the Environmental Health workforce

When the nature and scale of the COVID-19 problem became apparent, some countries lost no time in calling on EHPs to perform a wide range of activities, often developing and implementing novel interventions. Elsewhere, there is the suggestion that governments might have been ignorant of the Public Health resource available through their EH workforce, or otherwise reluctant to see them mobilised in the cause of public health.

It is hoped that these 'benefits' will become more evident through the publication of this book, but there remains the monetary cost, and the argument that in the UK central government receives the benefit of many thousands of EHPs trained and employed largely at the expense of others. . These costs are substantial when you aggregate fees due for theoretical instruction and accommodation, salary during a practical training placement, and professional assessment. It seems inequitable for some Public Health-related professions to see their passage to qualification set out and financed by the State, while would-be EHPs in many countries are left to self-finance, and local authorities to fund training through council taxation. Together, this places the financial burden on communities.

Recommendation: The costs of preparing an EHP for practice, from entry to higher education through to qualification, including registration where required, should be calculated, then, having considered and made deductions to take account of the benefits to all parties along the way, presented as an investment to central government, the cost of which should be borne by the state, or sponsored by other means. With such cost-benefit data to hand, this may serve as a means of comparing the traditional routes to qualification with alternative graduate-type schemes, taking into account the fact that EHPs are unlikely to be in receipt of high salaries on qualification, which itself is a deterrent to able individuals considering the profession.

8.13 Preventing the future eclipse of Environmental Health

We think Professor Winslow would be proud to know that his notion of the science and art of Public Health is now, 100 years on, so well understood and appreciated across the globe, though likely appalled at the circumstances for this. It is perhaps why, for example, in the UK, local authority Directors of Public Health have appeared so extensively in the media coverage of the pandemic, being praised for their commitment to their communities and their repeated argument for local intervention.

One of the most positive features of the pandemic up until now is how Public Health practice, or Environmental Health practice in its literal sense, has become key to breaking the routes of transmission of the virus through hand washing, physical distancing, disinfection of surfaces, adequate ventilation and the wearing of face coverings. It is unfortunate that, despite all the effort EHPs have put in to see these 'non-pharmaceutical measures' implemented, one hears them referred to in the media as mere 'public health officials'.

In countries where Public Health is traditionally practised in multidisciplinary teams, Environmental Health has no need to be identified, or EHPs, named, for their contribution to be recognised. However, in others, such as England and South Africa, where the practices of 'Public Health' and 'Environmental Health' have been structurally separated, or not fully reunified, names and titles matter. The situation has not been helped by the term 'Environmental Health' itself being dropped by some local authorities and subsumed into departments with wider functions and titles like 'Regulatory Services', thereby reinforcing the 'law enforcement' stereotype.

During times of crisis, names, titles and designations probably matter more than they would do normally, but there needs to be an agreed title that is uniform across the country, indeed throughout the world. In this way EHPs might never again be overlooked or side-lined, since 'out of sight' tends to be 'out of mind'.

Recommendation: Professional bodies and members across the world should lobby their governments to see their Environmental Health services, and those practising them, represented at all levels and tiers, and ultimately at the highest governmental level. Countries that have a 'Chief Environmental Health Officer' or 'National Director of Environmental Health' sitting alongside the chief medical officer speak of its importance and value. Based on the contribution made by EHPs during the pandemic to date, this is undoubtedly justified.

8.14 Is the EHP's 'health protection' role being missed in their title?

Critics of the term and title 'Environmental Health' have long since called for this to be changed in order to better reflect what those practising sought to achieve rather than what it embraced. Certainly, the use of 'Practitioner' in the title of those employed to deliver Environmental Health was arguably an improvement on 'Officer' in the UK by removing the distinction between those employed in local government and those in other sectors. In South Africa, for example, the title 'Practitioner' reflects the higher education qualifications of EHPs, when compared to other 'Officer' posts that require graduation from high school. As has been demonstrated here, there should be a profitable co-existence between practitioners from different sectors, all are bound by a common purpose and standards of professional conduct.

Recommendation: Seeking a name change name for those practising Environmental Health would be unwise for many reasons, not least because it has reacquired relevance and meaning through COVID-19. In addition, many are happy to be identified as 'Environmental Health Practitioners'. However, their closer ties to those with a different but complementary remit is an opportunity to ascribe the general purpose of Environmental Health practitioners as 'health protection', and, working together with other health professionals on tackling the COVID-19 pandemic and beyond, describe their common goal as 'disease prevention', and ultimately, 'health improvement'.

8.15 Enhancing the role of local EHPs in 'test-trace-isolate'

Arrangements to mobilise a 'Test and Trace' scheme should have featured heavily in every emergency response plan across the world. While other non-pharmaceutical interventions have been adopted to limit the onward transmission of the virus, the single most important intervention, by far, remains the capacity and willingness to implement swiftly and decisively systems for case finding, testing, contact tracing and isolation. Where this was practised assiduously it has worked spectacularly well for COVID-19, though with two vital additional components: imposing restrictions on people entering from abroad, and properly supporting those required to go into isolation.

The critical window of opportunity to make 'test, trace and isolate' work is clearly when numbers of confirmed cases are low enough for sufficient trained and experienced Public Health practitioners to contact every case and contact. By implication, then, if you have more suitable personnel to call on to interview confirmed cases and their contacts, then the greater the capacity of the system to put people safely into isolation. However, it is not just a question of the simple number of contact tracers available but their level of expertise. It is mistake to believe that it is easy and can be accomplished by a minimally trained, lay-person on the telephone. Securing the co-operation of, and accruing information from, a reluctant positive case, especially if they are poorly and frightened, is a genuine skill.

Many authoritative sources are available that describe how test-trace-isolate has been used to good effect, these covering different pathogens responsible for outbreaks, epidemics and pandemics down the years. They also reinforce the message that the means of delivering this most fundamental series of public health interventions must be put in place now, even while further pandemic waves are diverting attention to other measures. We would draw readers' attention to the Independent Scientific Advisory Group for Emergencies' 'blueprint to achieve an excellent Find, Test, Trace, Isolate and Support system' which calls for a response that is conducted at a local level since a virus that '...relies on families, groups and communities within which to spread' demands for is control and eventual elimination, key health professionals including GPs, health visitors and environmental health practitioners working together '...in and within those communities locally' (Independent SAGE, 2020)

Recommendation: In countries, regions and states where there is a reluctance to accept that making first contact with an index case or their contacts is a job best performed unscripted, by a locally based health professional with the skills possessed of a qualified EHP, then this should be picked up by the professional body and the case made repeatedly in whatever ear of governance is suggesting otherwise. Canvassing for better recognition of EHPs in matters to do with infectious disease control should begin now before the next wave of COVID-19 arrives, or whatever follows. This role should be built into emergency resilience plans at local and national level.

8.16 Recognising that the SARS-CoV-2 pandemic is not unprecedented, should not set 'lives against livelihoods', and is far from over

This is neither the time nor place to identify those who should be held responsible for the shortcomings in a country's response to COVID-19. However, when the causative agent was so similar to one dealt with effectively less than 20 years before, it is hard not to do so. When public health systems are unable to prevent healthcare services from being overwhelmed in weeks, with millions dying across the world, surely we have a duty to consider what went wrong and suggest why.

Perhaps the world's response since January 2020 boils down to the inability or unwillingness of nations to act fast, hard and with singular resolve. Some failed to heed the warnings from recent history and either put off constructing a functioning public health service, or worse, had such a service, but then chose to alter its functionality and/or failed to resource it adequately such that all but a handful of countries were primed and ready to act decisively.

EHPs have found themselves playing an important, sometimes vital, role in meeting the on-going and fearful threat to life and livelihoods caused by the coronavirus pandemic. In the main they have been directed to act by governments unable to set the economy aside and let public health advice truly decide policy. That governments were warned, sometimes repeatedly, of the consequences of delay and indecision as they sought to balance 'economic' and 'public health' interests suggests a false dichotomy:

> Trying to appease both public health demands and the libertarian views of the free market has led to not only astronomical death tolls, such as in the US, UK and Brazil, but in flailing economies.

Half-way compromises do not work in response to pandemics and have just dragged out the pandemic for all.

Wenham (2021)

EHPs around the world might reasonably come forward as reliable witnesses to any future inquiry into their governments' efforts to simultaneously protect 'lives' and 'livelihoods' since they have been on the front-line in seeking to achieve both, and arguably know the challenges better than any other health professional or politician.

At the time of writing, we watch and wait while governments do their best to make us safer through vaccination. However, to serve the purpose for which people yearn – a return to normality – there has to be an acceptance of the consequence of worthy sound bites like 'we are none of us safe until we are all safe', since this carries with it a duty to provide the benefits of vaccination to those who need it most and can least afford it. When the 'competition' is between those nations that can, and those that cannot, the challenge of vaccinating the world effectively may prove a task we measure in years, not months.

Recommendation: EHPs are equipped to recognise the early-warning signs of emergent infectious diseases and their wider public health consequences, including economic ones, and they should feel empowered to 'call this out'. Furthermore, whenever they hear claim of expertise in a matter to do with the behaviour of an infectious agent, or certainty as to how this might play out in the community, they should trust less and question more. This supports the bolstering of academic skills such as critical thinking within our academic EH programmes.

Further, until therapeutics and prophylaxis see the world through this crisis, it remains as important as ever that the 'non-pharmaceutical interventions' documented here continue if we are to limit the impact of the pandemic, both in terms of health (saving lives) and the economy (saving livelihoods), However, first and foremost, these must serve to protect the most vulnerable members in our society. As key players in the development, implementation and sustainability of NPIs, EHPs should become their natural advocate.

8.17 A 'one world, one health' vision for Environmental Health

While 'Think Global, Act Local' may now seem a tired slogan 30 years on from the Rio Earth Summit where it was coined, we were reminded of it when the president of the International Federation of Environmental Health, Professor Susana Paixão, chose to focus on Sustainable Development in her 'World Environmental Health Day' address on 26 September

2020. This prompted us to pick up on the notion – 'One World, One Health' – once considered an idea more suited to philosophical debate, but which, many believe, provides the means of framing the problems behind the world's ails, given their complexity and interconnectedness.

Beyond cruelly exposing weaknesses in how countries respond to infectious disease, COVID-19 has uncovered the frailty of national and global policy across an array of harms including deforestation, the uncontrolled disposal of waste, poor animal husbandry, insecure and unsustainable food chains and the climate emergency. These, taken together with other factors such as antibiotic resistance and indiscriminate use of artificial fertilisers, may have all played a part in creating the conditions that made the current pandemic inevitable.

If Environmental Health decides to commit to this 'bigger picture', it will be in good company, and need look no further than the WHO to bring 'One Health' a step closer to 'centre stage'. Those who heard or read the director-general of the WHO, Dr Tedros Adhanom Ghebreyesus's opening remarks to the World Health Assembly on 18 May 2020 will remember them as sobering but inspirational. They speak, in a political sense, to world leaders, but in the three passages assembled below, they might be considered the preface to a charter for the 'One Health' movement, and as a rallying point for all who hold Environmental Health dear.

> The pandemic is a reminder of the intimate and delicate relationship between people and planet. Any efforts to make our world safer are doomed to fail unless they address the critical interface between people and pathogens, and the existential threat of climate change that is making our earth less habitable.
>
> The world can no longer afford the short-term amnesia that has characterised its response to health security for too long. The time has come to weave together the disparate strands of global health security into an unbreakable chain – a comprehensive framework for epidemic and pandemic preparedness.
>
> COVID-19 is not just a global health emergency, it is a vivid demonstration of the fact that there is no health security without resilient health systems, or without addressing the social, economic, commercial and environmental determinants of health.

8.18 Final word

As we set out to do, this book is about EHPs by EHPs, for all to better appreciate what this remarkable and largely unsung cadre of health professionals have brought to the defence of the people of the world – living up to the Chartered Institute of Environmental Health's motto: 'Friends of the Human Race'.

References

Couch, R., Stewart, J., Barratt, C., Dhesi, S., & Page, A. (2012). *Evidence, Research and Publication: A Guide for Environmental Health Professionals.* Lulu publications eBook, download for free via: https://ukehrnet.wordpress.com/ehrnet-resources/ebook/

Independent Scientific Advisory Group for Emergencies (2020). The independent SAGE report 19. A blueprint to achieve an excellent Find, Test, Trace, Isolate and Support system. Available at: https://www.independentsage.org/wp-content/uploads/2020/10/New-FTTIS-System-final-06.50.pdf (accessed: 14 February 2021)

International Federation of Environmental Health (2020). Briefing by IFEH President, Professor Susana Paixão, December 2020 Newsletter. Available at: https://www.ifeh.org/docs/ifeh_newsletter/IFEH_Newsletter_Dec_2020.pdf (accessed: 20 February 2021)

Morse, T., Chidziwisano, K., Musoke, D., Beattie, T.K., & Mudaly, S. (2020). Environmental health practitioners: A key cadre in the control of COVID-19 in sub-Saharan Africa. *BMJ Global Health*, 5, e00314. https://doi.org/10.1136/bmjgh-2020-003314. Available at: https://gh.bmj.com/content/bmjgh/5/7/e003314.full.pdf (accessed: 11th February 2021).

Wenham, C. (2021). Editorial: What went wrong in the global governance of covid-19? *BMJ*, 372, n303 (Published 4 February 2021). Available at: https://www.bmj.com/content/372/bmj.n303 (accessed: 14 February 2021)

World Health Organization (2020). Opening remarks to the World Health Assembly on 18 May 2020 by the director-general (Dr Tedros Adhanom Ghebreyesus). Available at: https://www.who.int/director-general/speeches/detail/who-director-general-s-opening-remarks-at-the-world-health-assembly (accessed: 14 February 2021)

Index

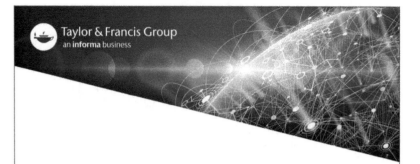